BISON
BOOKS

TEACHER IN SPACE

Christa McAuliffe and the Challenger Legacy

COLIN BURGESS

Foreword by
Grace George Corrigan

University of Nebraska Press
Lincoln and London

Library of Congress Cataloging-
in-Publication Data. Burgess,
Colin, 1947 – Teacher in space :
Christa McAuliffe and the
Challenger legacy / Colin
Burgess ; foreword by Grace
George Corrigan.   p.   cm.
Includes index.
ISBN 0-8032-6182-9 (pa : alk.
paper) 1. Challenger (Spacecraft)
–Accidents. 2. McAuliffe,
Christa, 1948–1986 – Influence.
3. Teachers – United States.
4. Space flight. I. Title. TL867.B85
2000 629.45'0092 – dc21
[B]   99-056385
By agreement with the author, all
royalties from this book will go
to the Christa McAuliffe Fund.

HIGH FLIGHT

Oh, I have slipped the surly bonds of earth,

And danced the skies on laughter-silvered wings;

Sunward I've climbed and joined the tumbling mirth

Of sun-split clouds – and done a hundred things

You have not dreamed of – wheeled and soared and swung

High in the sunlit silence. Hov'ring there,

I've chased the shouting wind along and flung

My eager craft through footless halls of air.

Up, up the long, delirious, burning blue

I've topped the wind-swept heights with easy grace,

Where never lark, or even eagle, flew;

And while with silent, lifting mind I've trod

The high, untrespassed sanctity of space,

Put out my hand, and touched the face of God.

John Gillespie Magee Jr., Canadian Air Force.
Killed in action, 1941, age nineteen.

Contents

Illustrations follow page 50

More than fourteen years have now passed since we lost Christa Corrigan McAuliffe on that terrible day in January 1986. I say we, because my daughter – a teacher – managed to touch the lives of many people and continues to do so. Today other teachers who found encouragement and inspiration in her story stand tall and are proud of their profession, while students tell me they are striving harder to reach and even exceed their personal goals. They know that Christa wasn't afraid to extend her own limitations, and whenever she did something she always did it well, to the best of her ability and unfailingly.

Christa was an innovative social sciences teacher who believed that the ordinary person contributed as much to history as did presidents and kings. She ran her classroom based on mutual respect, asking but two things of her students – that they be themselves and that they do their very best.

Being chosen to do an extraordinary thing did not alter who Christa was or what she valued. She welcomed her space flight as a grand opportunity for education and her year with NASA as a unique and positive means of drawing attention to schools, children, and community – a year of awareness and elevation of teachers.

The space frontier, like the future, belongs to all of us, and Christa felt that sharing her experiences would excite students and teachers. She not only was going on what she called the Ultimate Field Trip but intended to use the space program as a wonderful tool for education.

My daughter's dream did not die seventy-three seconds after liftoff. Her personal mission carried on, and her spirit lives throughout the world. Many productive tributes have been established, while schools and other public institutions bear her name. Educational opportunities and scholarships for the young and not so young have been created, while Challenger Learning Centers across the United States and Canada honor the 51-L crew of Dick, Mike, Judy, Ellison, Ron, Greg, and Christa.

The real heroes in life are people just like Christa – people such as parents and teachers, who help children in ways that they may never even realize. Real heroes are people who take everyday problems in their stride and persevere without ever losing sight of who they are and what is important.

Through author Colin Burgess, I know my daughter's message will reach many young people – not just in his homeland, Australia, but around the world, and for this I thank him. I have read some of Colin's other works and feel honored that he is now telling Christa's story.

Grace Corrigan

Acknowledgments

There are many to whom I owe sincere thanks for their assistance in the compilation of this book. First and foremost to Grace George Corrigan, who not only supplied the foreword but also checked the text to ensure complete authenticity and, in a touching display of trust, supplied the author with many original personal photographs for reproduction as illustrations. It was not only material help; Grace Corrigan lent her full support to this publication knowing it would reach and hopefully impact on the lives of many thousands of young people. If some of these readers are inspired to pursue their educational ambitions with a little more vigor or to achieve higher goals because of her story, then Christa McAuliffe and her dreams will live on.

Christa's backup, Barbara Morgan, also took the time to read the manuscript for this book and recommend some small but nevertheless vital changes. I am very grateful to Barbara, in the first place for her prestigious (and prodigious) assistance at short notice and secondly for her personal words of support and encouragement. They truly meant a great deal to me.

Endless thanks again to spaceflight researcher Kate Doolan for her immeasurable help in tracking down source material and to Jeff Green and Mike Ryan of the New Zealand Spaceflight Association for their ongoing and considerable support for this and other space projects. I am also indebted to the good folks at the Christa Corrigan McAuliffe Center in Framingham, Massachusetts, and Nancy Laprade at the Christa McAuliffe Planetarium and Foundation in Concord, New Hampshire, not only for their enthusiastic assistance but also for keeping alive the spirit and vitality of a beloved New Hampshire teacher whose life is commemorated in these wonderful places of excitement and learning.

Photographic material supplied by NASA and its use is gratefully acknowledged, and in this regard I am particularly thankful once again to the tireless Debbie Dodds and Jody Russell at the Johnson Space Center in Houston and to Margaret Persinger at the Kennedy Space Center, Florida.

Previously published sources of reference were the books *A Journal for Christa* (Grace G. Corrigan, University of Nebraska Press, 1993); *I Touch the Future* (Robert T. Hohler, Random House, 1986); *Challengers* (the staff of the *Washington Post*, Pocket Books, 1986), and *Judith Resnik: Challenger Astronaut* (Joanne E. Bernstein and Rose Blue with Alan Jay Gerber, Lodestar Books, 1990). I am also indebted to Barry DiGregorio for allowing me to quote from his article "The Time Is Right," copyright 1994 by Barry E. DiGregorio.

And finally, this book had its genesis in the unbounded curiosity of young students at schools I visited when giving talks on the history of space flight. Without exception they wanted to know much more about the teacher named Christa who would have flown in space. This book is for them and for young people everywhere.

Teacher in Space

*Space is for everybody. It's not just for
a select group of astronauts. That's our
new frontier out there.*

Launch Day

1

It was freezing cold on Launch Pad 39B. As night gave way to morning, a dull and wintry sunrise slowly spread warmthless daylight over Cape Canaveral and the Kennedy Space Center in Florida. Forty-five massive floodlamps lighting the steel launch tower were turned off, and a gleaming space shuttle stood poised for its next mission.

The space shuttle, or orbiter, was named *Challenger*, and like three other winged spacecraft in the fleet was named for a famous ship of exploration. This was to be *Challenger's* tenth flight into space, on a mission carrying seven astronauts – five men and two women.

There had been several delays to the flight, the last just a day earlier. Despite the extreme chill the cape forecast was for clear skies, and optimism ran high. The date was Tuesday, 28 January 1986.

At seven o'clock that morning the temperature stood at minus five degrees Celsius. NASA's "ice team" had just returned from their latest inspection of the shuttle's huge external fuel tank. The tank had been refueled with supercooled propellants, causing ice to crust on its exterior. It was the team's job to check the amount of ice at the launch pad in case it reached dangerous levels.

Their report was alarming. Sheets of ice and even thick stalactites had formed on many of the launch pad facilities. With the massive vibration of takeoff this ice could be dislodged and tumble down onto the ascending spacecraft. The team managed to break off and remove many of the thicker sections of ice, but it soon built up once again in the chill, moist air. The situation was far from ideal.

When she was woken at 6:20 that morning, Christa McAuliffe's first thought was to peer excitedly through her window at the distant launch pad. She never grew tired of the sight before her. Brilliantly illuminated by floodlamps, *Challenger* pointed at the early dawn sky, mounted on its massive rust-colored external tank, which in turn was flanked by two gleaming white boosters.

The external tank would feed liquid oxygen and liquid hydrogen through two fuel lines into the orbiter. There they would mix and ignite, powering the spacecraft's rear-mounted main engines. Adding further massive thrust to the launch, the two booster rockets carried more than a million kilograms of solid fuel propellant. When the countdown reached zero they would ignite and keep firing until the fuel was spent. They were very similar in principle to the skyrockets Christa and her friends had sent aloft in their childhood.

Dressed casually, the crew met in the dining area for NASA's traditional prelaunch breakfast. In the center of their table sat a huge cake, decorated with a design of *Challenger* and the names of the crew. After they'd enjoyed breakfast together the astronauts took an elevator back up to their rooms, where they would partially suit up for the flight. Following this they reassembled in the crew quarters. Here they received their final weather briefing and were told about the low temperature and ice at the launch pad. Despite this it was a glorious day, with weak winds and increasingly blue skies.

Outside, in the freezing weather, invited guests sat and shivered in their special viewing area located a safe distance from the pad. Relatives of the crew, reporters and cameramen, special guests, and a large group of school children sat huddled against the cold, hoping that this would be the day *Challenger* and its crew finally rocketed skyward and went into orbit. The flight had already been delayed several times due to malfunctions and bad weather. Just a day earlier the countdown had reached T−9 minutes before a faulty bolt in the external hatch handle jammed, putting the launch on hold. It took four long hours to repair the fault, by which time strong winds had swept into the cape and the crew were told the launch was off once again. For the past thirty-eight days *Challenger* had stood patiently on its launch pad, and now everyone prayed for a successful launch.

Striding out to their transfer bus for the trip to the launch pad, the seven astronauts waved at friends, staff, and media representatives along the way. They clambered into the bus and just before eight o'clock were once again on their way out to Launch Pad 39B.

On arrival at the pad they disembarked and entered a small lift, which took them up to a long, canvas-covered walkway on the same level as the shuttle's open hatch. This walkway opens out to an area known as the White Room, where shuttle crews are furnished with the rest of their flight suits and safety harnesses. They all managed to joke with members of the launch pad crew, and Christa commented to one that it was a great day for flying. "We are going to go today, aren't we?" she added.

The two pilots, Dick Scobee and Mike Smith, donned their helmets and escape harnesses and were assisted into the shuttle through the round hatch. Then it was the turn of the two mission specialists who would sit behind them on the flight deck, Judy Resnik and Ellison Onizuka. Occupying seats on the mid-deck for takeoff, Greg Jarvis, Christa McAuliffe, and Ron McNair were next to clamber aboard.

Just as she was preparing to enter *Challenger* through the crew hatch a launch technician handed Christa a last-minute present that caused her to laugh out loud in appreciation. It was a shiny red apple, for America's favorite teacher. Although Christa could not take it with her she thanked the man for his kindness and returned the apple, with a request that he give it to her on the crew's return from space. The good-luck present caused Christa to enter the hatch with a happy smile on her face.

On 19 July 1985, just six months earlier, Christa McAuliffe had stood in the Roosevelt Room at the White House in Washington DC, fighting back tears as she clutched a small bronze statue. The statue was of two people – a teacher and a student. The teacher was pointing upwards, toward the stars, and the student was also looking to the heavens, following his teacher's gaze. Vice President George Bush smiled and shook Christa's free hand, and she gulped nervously as she leaned toward the microphone. She was aware that millions of people across America were watching the live telecast, and they were waiting to hear from the person chosen to be the nation's first teacher in space. For a moment her mind went blank at the thought of the incredible task that had been entrusted to her.

"It's not often that a teacher is at a loss for words," she finally stammered. "I know my students wouldn't think so." The other teacher finalists standing behind Christa smiled as she continued. "I've made nine wonderful friends over the last two weeks, and when that shuttle goes, there might be one body" – she choked back tears and pressed a finger to her lips as she recovered her composure – "but there's going to be ten souls I'm taking with me."

Everyone applauded, and a few moments later Christa was ushered out of the presentation room to a lawn on the grounds of the White House where an eager throng of reporters and photographers were waiting for her. "How does it feel?" one called out as she moved to the battery of microphones. Christa laughed; she would have to get used to these questions now that she had suddenly become her nation's newest hero. "I'm still kind of floating," she answered. "I don't know when I'll come down to Earth."

2

When I was in junior high school I remember somebody sitting down and talking about career choices and saying I could be a nurse, a secretary, a teacher, or an airline stewardess – nothing else.

Sharon Christa McAuliffe was born in Boston, Massachusetts, on 2 September 1948. At that time her parents, Ed and Grace Corrigan, lived in a tiny apartment near Fenway Park, home of the Boston Red Sox baseball team. They had been married only fourteen months when their daughter was born, but they had known each other since their high school days and were always considered an inseparable couple.

At the time the Corrigans learned they were to become parents, they discussed possible names for their baby and finally decided on Christopher, Ed's middle name, if they had a son. The birth of a daughter caused a small but pleasant setback to these plans, and they soon chose to name her Sharon, with a middle name of Christa – a form of Christopher.

Eventually the new parents began to call their bouncing baby daughter Christa, and the name just stuck, according to her mother, Grace. "For six months we called our baby Sheri Christa. She was very blond, and we became fond of the name Christa, thinking somehow that it really suited her better. So we just dropped the name Sharon. It took a little while, but soon everyone was using the name Christa. We never thought of her as anything but Christa after that."

In later years their daughter would sign her name S. Christa Corrigan, and she continued to place the "S" before her name in the thousands of autographs she gave prior to her flight.

As a baby Christa suffered from asthmatic bronchitis, and when she was just six months old she was hospitalized with chronic intestinal problems. The doctors had to shave her head on both sides and insert tubes into her scalp, through which she was fed a mixture of glucose and water. It was a difficult time for the Corrigans and their little daughter, but five weeks later she had recovered sufficiently to go home.

Ed Corrigan was still studying at Boston College and consequently had very little money to put down on a home, but he had served as a navy radioman during the war and was able to apply for veterans' housing. Eventually they were offered a converted barracks building in a Boston public housing project. It wasn't a palace by any means, but the Corrigans were thrilled to be in their own home at last and set about painting and decorating.

Little Christa now had her own bedroom, where she and her puppy Teddy would play while her mother painted and scraped the walls. From a very early age she loved listening to music, and her father would later describe her as a "nonsleeper." At the age of two, Christa would clamber out of bed in the middle of the night and listen to records, singing and dancing along to the tunes.

Her spirit of adventure was also becoming evident. One day she hopped on her tricycle and began pedaling along busy Columbia Road, which led to the city of Boston. It was her dog Teddy who prevented her from going too far, running onto the road and circling the tricycle, barking as hard as he could until the cars stopped and Christa was rescued by her anxious parents. "We never did figure out where she was going," her father said later.

Ed Corrigan earned his degree in industrial management just a week after the birth of Christa's baby brother, named Christopher. To avoid any confusion with his sister he quickly became known to everyone as Kit.

There would eventually be three more children: Steven, Lisa, and Betsy. The family moved to Framingham, west of Boston, where Ed took an executive job with a large chain of department stores, while Grace involved herself in community and church work, becoming well respected for her hard work and cheerful disposition.

As Christa grew up she remained chubby-faced, and her impish grin would reveal two prominent front teeth. Some of the boys at school teased her because of this, calling her "Chipmunk." However, family photos of Christa at this time show a pleasant young girl with a charming smile, bright eyes, and auburn hair, which would slowly darken over the years. She even attended modeling school for a short time and appeared on a television fashion show.

As the oldest of five children, Christa took a great deal of responsibility for their upbringing. She would patiently explain how things worked, tell bedtime stories, take over when her parents went out, and generally act as a caring big sister. Any time she returned from her grandmother's house or a camp she would bring small gifts for her younger brothers and sisters. It might be only a small piece of gum or a flower, but she always made sure everyone got something.

In time she took piano and dance lessons and even taught her younger sisters how to sew. On her sixteenth birthday Christa was given a sewing machine, and soon she was turning out beautiful dresses and jackets with competent ease. She also enjoyed participating in Girl Scouts, especially when they went on field excursions. With all these activities to occupy her time, she found it difficult to keep up with piano lessons and soon switched to playing the acoustic guitar. Once she had mastered a few chords she began teaching songs to her Girl Scout troop, accompanying herself on the guitar.

On 5 May 1961, still in the seventh grade at school, Christa and her classmates watched intently as a historic event unfolded on the school's black-and-white television. On that momentous day Alan Shepard became America's first man in space, riding to glory atop a Redstone rocket. It was only a fifteen-minute suborbital flight, which meant Shepard was fired into space and returned to an Atlantic splashdown nearly five hundred kilometers downrange, but the United States had now entered the so-called space race with the Soviet Union. Christa would later recall her fascination with those early space launches. "I remember the excitement in my home when the first satellites were launched. My parents were amazed, and I was caught up in their wonder. In school, my classes would gather around the TV and try to follow the rocket as it seemed to jump all over the screen. I remember when Alan Shepard made his historic flight – not even an orbit – and I was thrilled!"

According to her mother Christa had an active and relatively normal life when she went into junior high school at Boston's Marian High, a coeducational Catholic school.

She did well in class and got along with her teachers and classmates, enjoying her extracurricular activities as well. As a junior she played on the Marian High School junior varsity girls' basketball team and also was the pitcher for St. Jeremiah's softball team.

Barbara [Eldridge] and Christa were sitting together in the bus going to school February 21, 1962. The day before Marian High had canceled school so everyone could watch while John Glenn made his historic flight on *Friendship 7*, orbiting the earth three times before landing safely in the Atlantic Ocean. They discussed the wonder of the flight, and Christa said to Barbara, "Do you realize that someday people will be going to the moon? Maybe even taking a bus, and I want to do that!"

One of her teachers at Marian High, Sister Mary Denisita, remembered Christa Corrigan as a keen student in a highly regarded college preparatory school of above average students. "Her face was very alive, very interested," Sister Mary told Christa's biographer, Robert Hohler. "You could tell by looking at her that she was excited about everything life held before her."

The personable young girl was always eager to learn and study, with a feisty determination to do the best she could at all times, and this led to her achieving a place in the school's National Honor Society. Christa's senior-class yearbook noted that her activities included the student council, glee club, orchestra, girls' basketball, ceramics, German club, and the senior play.

While at high school, Christa caught the eye of a fellow student named Steven James McAuliffe, who finally plucked up the courage to introduce himself. They were soon in each other's company at every opportunity, and a youthful romance blossomed. When they were both sixteen Steve decided Christa was the girl for him and proposed. To his relief and joy she accepted, but they wisely realized that any wedding would have to wait until Steve completed college.

Following their graduation from high school Steve attended the Virginia Military Institute. Meanwhile Christa had decided on a career in teaching and, having discussed various college options with her father, enrolled at the hometown Framingham State College.

At college Christa took a course called "The American Frontier," conducted by history professor Carolla Hagland. She would later tell the professor that his course on women pioneers and their fascinating diaries of life on the frontier had proved to be of immense interest and had forever changed her life.

Other more contemporary matters were also having an impact on the young college student. America was heavily enmeshed in the Vietnam War. Robert Kennedy and Martin Luther King Jr. had been assassinated while the country was still coming to terms with the earlier slaying of President John Kennedy in 1963. Christa wanted to register her protest and sorrow at her country's involvement in Vietnam but not to the extent of organizing rallies or burning the American flag. Instead, like many other college students, she made a passive protest by wearing a black armband to her graduation ceremonies.

Christa graduated with a bachelor of arts degree in the spring of 1970, and her marriage to Steve took place on 23 August that year. The happy newlyweds moved into an apartment in Virginia, Maryland,

and, while Steve attended Georgetown University Law School, Christa found part-time employment teaching American history at several junior high schools. A temporary job as a waitress in a Howard Johnson's restaurant helped pay the bills. From 1971 to 1978 she taught American history and English at Thomas Johnson Junior High School in Lanham, Maryland. Meanwhile Steve had graduated from Georgetown and worked for a time as a defense attorney with the Judge Advocate General's Corps, the army's law firm. In 1975 he became a law clerk in a private practice with the president of the Virginia State Senate.

On 11 September 1976 their first child, Scott Corrigan McAuliffe, was born, and Steve began planning a more permanent home for his family. He wanted to work in Washington, but Christa had other ideas. When she had camped with the Girl Scouts in New Hampshire and skied at resorts not far from the Corrigan's home in Framingham, she had fallen in love with the friendliness of the New England people and the rustic serenity of New Hampshire. A recent trip to visit friends had only reinforced her desire to make a home in the area.

Steve was soon offered a good position with the Justice Department in Washington, promising excellent career prospects, but Christa remained adamant. "You can live where you want," she said, "but Scott and I will live in New Hampshire." Despite Steve's pleas that they would both enjoy better career opportunities in Washington, Christa was unswayed. Fortunately Steve was offered a good position with the attorney general's office in Concord, New Hampshire, and became an assistant in the civil division. It was a good enough stepping-stone to a secure future, and he was happy with the position. Christa was relieved and delighted for her husband.

With their careers now on the move they could begin searching for a new home, and finally the right one came along. It was a quaint, three-story, brown-shingled Victorian house in Concord, and they decided they could not do better at the time. The house needed quite a lot of work, but it was situated opposite a small, leafy park, which they felt was a perfect setting. Concord itself was a peaceful town of around thirty thousand people, just one hundred kilometers north-west of Boston.

Two years after Scott was born Christa began teaching English and history at Rundlett Junior High School in Concord, but nine months later she was reluctantly laid off due to severe budget cuts. By that time, however, she had earned her master's degree in educational administration at Bowie State College, completing a thesis on the acceptance of handicapped children into regular classrooms.

A second child came along on 24 August 1979, this time a little girl they named Caroline, after Christa's much-loved Aunt Carrie. Soon after the happy event Christa was back teaching, this time at Bow Memorial School.

In 1982 Christa was offered a new job teaching social studies at Concord High School, just a short walk from her home. She couldn't believe her luck, as Concord High was regarded as one of the state's best and most progressive public schools, with excellent teaching facilities. Steve, meanwhile, had joined a prominent law firm in Concord, Scott had begun kindergarten, and the family was settling into a comfortable life with a secure future in store.

Like her mother, Christa had become deeply involved in local community activities, raising money for the YMCA and the local hospital, acting in amateur presentations, jogging, playing tennis, and later playing volleyball in a team with Steve. When Scott began attending Sunday school at St Peter's Church she helped out by giving the children classes in religion.

Every year the McAuliffes would welcome an underprivileged, inner-city child into their home for a time as part of a caring program called A Better Chance. Steve sometimes felt Christa was taking on far too much, but he figured it was what she wanted. He knew his wife would not be happy unless she was out there doing her best to help others.

There is always something to learn,
always a way to make a lesson better.

3

Reveling in her position at Concord High, Christa taught American history, law, and economics, mostly in classroom 305. She enjoyed the teaching experience and loved involving her pupils in the process of their own education. At times they would dress up in period costumes when studying certain aspects of American history, and as junior historians they were encouraged to keep journals of their investigations.

As part of her social studies classes Christa brought prominent Concord townspeople into the classroom to talk about their work. Her students also kept an eye on stock-market fluctuations and interviewed local stockbrokers as part of their curriculum. One of her students from those days was Susan Dill:

I took her two Law for Young People classes. But we called them "Street Law" classes because she taught us what we should know when we got out on our own: how to stand up for our rights, all about landlords . . . and what happens in the courts to drunk drivers. She'd take us to the city hall and the courthouse, all thirty of us, to listen to the judge and lawyers and policemen.

She'd even bring in a car salesman to explain how you can get ripped off. Most of it stuck because what really mattered to her was us as people.

Favorite among Christa's educational activities were her famous field trips, which provided students with enjoyable and instructional excursions. In those days such outside activities were considered quite novel, but it was just her way of educating. She saw positive benefits in periodically removing students from stuffy classroom environments and familiar distractions.

This innovative approach to teaching certainly brought her to the attention of the school's faculty, many of whom did not quite share Christa's enthusiasm for outside excursions. Conversely, others came to admire the spirited young woman for her educational programs and noted the enthusiasm and correspondingly high grades of those in her classes. Christa, it seemed, had truly found her vocation.

In her role as a social historian Christa set up a course on a subject that had fascinated her since college – The American Woman. The emphasis of this course was the role women had played in raising families while their husbands were off fighting in different wars. In developing the course she discovered a wealth of remarkable material in the diaries and letters of pioneering women. As she later told an interviewer, "They didn't have a camera. They described things in vivid detail, in word pictures. They were concerned with daily tasks and the interaction between people, with hopes and fears. Those diaries and journals are the richest part of the history of our westward expansion. Without them, it would just be how many Indians were killed and the number of settlements started."

Christa's enthusiasm for the subject was infectious and would later cause one of her pupils to write, "Mrs. McAuliffe's course on the American woman changed my outlook on life. It was like she discovered something new every day, and she was so excited about it that it got the rest of us excited too!"

By August 1984 the McAuliffe family had settled quite happily into their New Hampshire lifestyle, although at times Steve still mildly reproached his wife for her busy schedule of work-related, social, and charitable activities. Christa, however, would not have it any other way; she loved being active in all aspects of life in Concord. Their home was always open to her students if they wished to discuss any scholastic or personal problems. It was not unusual for late-night visitors to ring their doorbell with some pizza and a pressing need to talk over something.

On 27 August that year, life in the McAuliffe household suddenly took a new and dramatic turn. That evening, as Christa and Steve were driving through Concord, an interesting bulletin on the car radio caught their attention. President Reagan was on the air announcing that a private citizen would be chosen to make a flight into space. That person, he declared, would be a teacher.

Christa turned to Steve, and he grimaced before directing his attention back to the traffic. He had seen that look on his wife's face before, and he knew without asking that she would be applying for the job. He sighed, shook his head, and then glanced at her again. She was waiting quietly for his approval – some sign that he wanted this new adventure as much as she did. He let a few moments pass before he relaxed and smiled at her. "Go for it!" he said.

In 1975, six years before the maiden flight of a space shuttle, NASA concluded that qualified nonpilots would be needed aboard their new reusable spacecraft. Pilots would always be essential to fly the shuttle, but specialists in diverse medical and scientific fields could conduct vital and intricate experiments while in orbit.

The physical requirements for these people would be far less stringent than those demanded during the early days of space flight, when only the country's best military pilots were chosen to fly the complex capsules that took America into space and onto the Moon. It was now time to reap the benefits of those pioneering flights through working and experimenting in space, which meant sending up specialists in a variety of scientific and medical fields.

NASA's Office of Life Science established the medical criteria for selecting these people, who had to be prepared to live and work in space for up to thirty days. Those chosen would earn their place aboard the shuttles by supervising the deployment of satellites and servicing faulty ones in orbit. They would also conduct onboard experiments for a variety of paying customers and carry out photographic and radar imaging studies of the Earth, its environment, and its resources.

A relaxation of the strict medical standards was possible, as the forces of acceleration during a shuttle launch would only be three times that of Earth's gravity – far less than those experienced by earlier astronauts. For the first time it would also mean that women could be considered for space flight. In centrifuge tests carried out with twelve women at NASA's Ames Research Center in California in 1973 no physiological problems had been identified.

The first astronaut group to include the new category of mission specialist or nonpilot astronaut was announced in January 1978. Six women were among those chosen, qualifying as NASA's first female space explorers.

Around this time a prototype space shuttle named *Enterprise* was undergoing tests to determine the shuttle's flight and landing characteristics. It had been constructed without rocket engines and was never designed to fly into space. After a piggyback ride into the California skies atop a specially modified 747, *Enterprise* would be released to glide back to a planned landing at Edwards Air Force Base. Following several flights this test series was deemed successful and concluded. Unlike earlier programs there would be no unmanned orbital test flights of the shuttle; the first time the winged spacecraft soared into space it would have a crew of two on board.

On 12 April 1981 the space shuttle program roared into life with a picture-perfect launch of shuttle *Columbia* from the Kennedy Space Center, carrying astronauts John Young and Bob Crippen on a successful thirty-six-orbit proving flight. By coincidence the launch of *Columbia* fell on the twentieth anniversary of the historic first space flight by cosmonaut Yuri Gagarin of the Soviet Union.

As the shuttle program evolved beyond its orbital test phase the names of the first mission specialists began to appear on crew manifests. Once in space they conducted medical and scientific experiments, leading to some remarkable discoveries about the human body in microgravity, the processing of new and unique alloys and chemicals, and the geological history and changing face of the Earth below.

Another category of astronaut was created – that of payload specialist. Although NASA undertook their mission training, these people were not members of the agency's astronaut corps. Instead they were representatives of various companies, the military, or nations with a specific job to perform in space. They oversaw the deployment of satellites from the shuttle's payload bay or conducted experiments that would have benefits for people back on Earth. Australian-born navy oceanographer Dr. Paul Scully-Power, for example, was a payload specialist on the thirteenth space shuttle mission in 1984. His observations from shuttle *Challenger* reshaped our understanding of the influence the oceans' currents have on our weather and the causes of such phenomena as earthquakes and droughts. During the mission he occupied the seat on *Challenger* later assigned to Christa McAuliffe.

In 1983 and with the apparent success of the space shuttle program, a task force of the NASA Advisory Council carried out an extensive study on the issue of allowing private citizens to fly into space. In its report to NASA administrator James Beggs it was recommended in part that "NASA should take the next step in opening space flight to all people by flying observers for the reason of meeting the purposes of the Space Act."

The report also set up critical guidelines for applicants, such as health, training, and space adaptability. The Advisory Council's intention was to provide a series of flight opportunities, "with each having a different objective or purpose and each aimed at a different occupation or group of people. Over the long term a broad cross-section of society would be included."

The genesis of the Teacher in Space program was now firmly in place, and it became official on 27 August 1984. On that day President Ronald Reagan made a stirring campaign speech to the nation, endorsing the recommendation of the Advisory Council and setting an epic challenge before the public.

It's long been a goal of our space shuttle, the program, to someday carry private citizens into space. Until now, we hadn't decided who the first citizen passenger would be. But today, I'm directing NASA to begin a search in all of our elementary and secondary schools, and to choose as the first citizen passenger in the history of our space program one of America's finest – a teacher. When that shuttle takes off, all of America will be reminded of the crucial role that teachers and education play in the life of our nation. I can't think of a better lesson for our children and our country.

Eligibility requirements were outlined in an application form, and well over forty-five thousand teachers requested a copy. The position was open to elementary- and secondary-level teachers in all public and nonpublic schools in the United States and its territories, in Department of Defense and State overseas schools, and in the Bureau of Indian Affairs. An applicant had to be an American citizen, have been a full-time teacher for five years, meet certain medical standards, be available for preflight training, and be willing to work full-time with NASA for one year.

Christa and several of her colleagues picked up their application forms during the annual conference of the National Council for the Social Studies, for which she was a New Hampshire delegate. She and Steve read it through when she got home and quickly realized that a lot of thought and effort would have to go into her responses to the questions.

Over eleven thousand teachers had their applications in by the deadline, although Christa, in her usual unhurried way, did not post hers until the first of February – the last day on which applications could be mailed. Completing the form had not been a simple matter; it required very detailed information from applicants, asking why they had applied and what they hoped to achieve if selected. They were set the task of writing several pages on each of the following topics:

Describe your professional development activities.

Describe your communication skills.

Describe your community involvement.

Describe your philosophy of teaching.

How do you help your students to develop a national and international awareness?

Why do you want to be the first U.S. private citizen in space?

Describe the project you would like to conduct during the mission.

How do you expect to communicate your experience during the year following your return?

Every application was judged by the Council of Chief State School Officers (ccsso), a professional organization made up of public officials responsible for education in all fifty states. They were looking for creativity and originality in each teacher's response, good communication skills, a firm commitment to education, and a proven community involvement.

The task of the ccsso was a mammoth one; they had to pare the list dramatically and select a first group of finalists comprising two teachers from each state. This list of finalists had to be ready for release to the media on the first of May.

Christa had put a considerable amount of effort into her application, but she knew there would be thousands applying, many of whom she felt would be better qualified and would put forward far better submissions. As always, though, she knuckled down to the task and presented her thoughts on why she wanted to be the first teacher in space. Selected extracts from her application give an insight into her reasons for wanting to fly aboard the shuttle and reveal her proposal for an educational program while on orbit.

As a woman I have been envious of those men who could participate in the space program and who were encouraged to excel in areas of math and science. I felt that women had indeed been left outside of one of the most exciting careers available. When Sally Ride and the other women began to train as astronauts, I could look among my students and see ahead of them an ever-increasing list of opportunities.

I cannot join the space program and restart my life as an astronaut, but this opportunity to connect my abilities as an educator with my interests in history and space is a unique opportunity to fulfil my early fantasies. I watched the Space Age being born, and I would like to participate.

In developing my course, The American Woman, I have discovered that much information about the social history of the United States has been found in diaries, travel accounts, and personal letters. This social history of the common people, joined with our military, political, and economic history, gives my students an awareness of what the whole society was doing at a particular time in history. They get the complete story. Just as the pioneer travelers of the Conestoga wagon days kept personal diaries, I, as a pioneer space traveler, would do the same.

My journal would be a trilogy. I would like to begin it at the point of selection through the training for the program. The second part would cover the actual flight. Part three would cover my thoughts and reaction after my return.

My perceptions as a nonastronaut would help complete and humanize the technology of the Space Age. Future historians would use my eyewitness accounts to help in their studies of the impact of the Space Age on the general population.

As the weeks passed and April finally swept in, Christa tried to avoid dwelling on her chances. There were times when she wondered if she could have expressed her thoughts a little better, but she need not have worried. On 16 April seven New Hampshire teachers were summoned for a further interview at nearby Walker Elementary School. Christa was one of the seven; soon their number would be two.

On 3 May 1985 the ccsso announced the names of the 114 elementary- and secondary-level teachers who had been chosen as state nominees in the nasa Teacher in Space program. Earlier, and to Christa's great astonishment, she had received a phone call from the chairman of selectors for New Hampshire, Dr. William Ewert, to say she had been selected as one of this group of finalists. She and Robert Veilleux, an astronomy and biology teacher, were the two chosen from the state of New Hampshire.

The executive director of the ccsso, Dr. William Pierce, said the application and selection process had gone better than anyone had anticipated. "It is a great pleasure for the council to be involved in this historic project," he declared. "The caliber of the applications from teachers throughout the country has been truly impressive. If their applications are any indication, we can be proud of the quality of teaching that occurs in the classrooms of the elementary and secondary school teachers who applied for this unique educational opportunity."

The next phase of the selection process for both state applicants required them to attend a medical examination and, within two weeks, prepare for a taped interview that would be forwarded to Washington for judging. It was a nervous time for Christa; she didn't feel the medical would present any problems, but she hadn't been in front of a camera since the age of four, when she had been interviewed on a local TV station. Fortunately a friend offered to help by taping Christa during several rehearsals, but she still felt far from confident.

The medical, as expected, went well, and to Christa's surprise she carried off the taped interview quite comfortably. Both finalists had to answer three questions and were given the first two in advance so they could prepare suitable answers. The third question would be asked live, without any chance to rehearse. In answering the first two questions Christa talked over some project ideas she had for the flight and how she felt it might affect her life.

The third question was far more profound: Dr. William Ewert asked Christa to describe her philosophy of living. She repeated the question slowly, at the same time mentally assembling an appropriate response. Then she looked around the studio, just as she would if giving a class lecture and launched into her reply.

Well, I think my philosophy of living is to get as much out of life as possible. I've always been the type of person who is very flexible and has tried new things. I feel that you need a little bit of organization, but I also think it's important to connect with people. I think the reason I went into teaching was because I wanted to make an impact on people and to have that impact on myself. I think I learn sometimes as much from my students as they learn from me. Being in an educational field has also been the type of thing where I can go out and see people later on in the community, see them in their twenties and having their families and growing up, and that has given me a good feeling.

My philosophy of living, I suppose, is to enjoy life and certainly to involve other people in that enjoyment, but also to be a participant and to enjoy all the things we have in this country.

On 22 June all 114 state finalists arrived in Washington DC for a five-day National Awards Conference that focused on various aspects of space education. Their ages ranged from twenty-seven to sixty-five, and they represented America's teaching community. Soon this number would be whittled down to 10, at which time the names would go to NASA, who would then take over responsibility for the final selection of the prime and backup candidates. However, in recognizing the enthusiasm and expertise of the group now assembled, NASA decided to offer all the finalists an additional challenge despite the fact that only 1 would be chosen to make the flight.

In a rousing speech at the opening banquet for the awards conference, the agency's administrator James Beggs acknowledged the skills and qualifications of the state finalists and gave the following assurance.

You are all winners, whether you fly or not. That is why tonight we are appointing you all ambassadors from the space program to your states, your districts, your schools, at home and abroad – wherever you can reach young minds. As our ambassadors to young America, we hope you will spread the word that space has become not just a place to visit, but a place where we are learning to do new, exciting, and useful things to benefit life on Earth. We hope you will share with our young people the opportunities and promise of the space program, and the excitement we all feel as we move to push back this new frontier with a permanent presence in space.

Later that evening the finalists received an enthralling briefing on the objectives of the mission in which the teacher would take part. The speaker was astronaut Mike Smith, who would be the pilot on that particular flight.

In order to reduce the field to just ten the ccsso had put together a National Review Panel comprising several distinguished individuals from education, business, science, and even the entertainment industry. Veteran space explorers "Deke" Slayton, Ed Gibson, Gene Cernan, and Harrison Schmitt represented the astronaut corps.

All nominees were extensively interviewed, both officially and informally, while their prerecorded interviews were screened and carefully reviewed. The judges then withdrew and met as a group to consider each applicant's claim for the prized position.

The finalists understandably became edgy as the time for the decision drew closer, but they were also enjoying the many privileges extended to them during the selection process. They spent an engrossing afternoon at Washington's National Air and Space Museum watching the remarkable imax film *The Dream is Alive* on a colossal screen nearly six stories high. The forty-minute film, shot over three shuttle missions, had a tremendous impact on the finalists. The huge screen and dynamic hi-fi soundtrack combined to send them on a realistic shuttle journey into space and gave them insight into life aboard the spacecraft.

The teachers were then introduced to several astronauts whose exploits had featured in the IMAX film, including mission specialist Judy Resnik, another of those scheduled to fly on the shuttle that would take one of them into space. In a rousing speech she echoed the words of James Beggs. "It's a shame every one of you can't be taken, because you're all winners. I sure look forward to meeting the lucky person, and with any luck I'll still be on that mission, too."

At three o'clock on the morning of 28 June the telephone began to ring in the McAuliffe house. Steve groaned and threw on his dressing gown as he trudged down the stairs, hoping it was not just another reporter or space enthusiast who wanted to speak to his wife. There'd been many such calls from well-meaning people, but some folks had no appreciation of time changes across the United States.

Christa could hear her husband muttering something over the phone and then listened as he tramped up the stairs. He turned on the light and sat on the edge of the bed, looking at her in a curious way. "It's for you," he announced calmly. "Someone from NASA. You made the final ten!"

*It didn't really matter how old you were
or what you weighed or how you looked.
They wanted a schoolteacher. Period.*

4

On 7 July, just six days after the official announcement of their selection, the ten finalists traveled to the Johnson Space Center in Houston for an exhaustive weeklong session of thorough medical examinations and briefings about space flight.

Christa had come to know the other finalists well, and she realized she was in very distinguished company. Many had academic qualifications as long as their arms, while some were already proficient pilots and scuba divers. The more Christa came to know them, the more she felt that the chance to fly in space would go elsewhere. She did not despair but set out to enjoy what remained of this incredible experience and the company of her colleagues.

There was Kathleen Anne Beres, who taught biology at Baltimore's Kenwood High School in Maryland, and Judy Garcia, who was due to begin teaching French and Spanish at Fairfax County's new Thomas Jefferson High School for Science and Technology in Virginia. Dick Methia was an eloquent English teacher at New Bedford High School in Massachusetts, while Bob Foerster, who designed software programs, was a sixth-grade math and science teacher at Cumberland Elementary School in West Lafayette, Indiana.

Peggy Lathlaen was responsible for educating gifted children at Westwood Elementary School in the Texas suburb of Friendswood, while Dave Marquart taught business and computer science at Boise High School in Idaho. Mike Metcalf was a former Air Force pilot and now taught geography at Hazen Union School in Hardwick, Vermont. Niki Wenger had been teaching for only five years but was enjoying her role as an educator for gifted children at Vandevender Junior High School in Parkersburg, West Virginia.

The other finalist in the group was Barbara Morgan, a dedicated second-grade teacher at McCall-Donnelly Elementary School in McCall, Idaho. Without any doubt they were an impressive group of talented educators from across the United States.

At the Johnson Space Center a final elimination process was begun, and the ten participants were put through several medical and psychological evaluation exercises. The first of these involved giving numerous blood and urine samples, being x-rayed from top to bottom, having their eyes and teeth checked, and answering hundreds of questions about their medical history.

One of the psychological tests that Christa truly dreaded was being placed inside a soft-skinned, zippered ball about three feet in diameter. It was being developed as a possible means of transferring trapped astronauts onto rescue vehicles, but for now it came in handy as a way of evaluating a test subject's susceptibility to claustrophobia. The occupant would be sitting in complete darkness with no idea of how long he or she would be there, and the thought of being sealed up in this way filled Christa with dread. She would later say that as she was placed into the ball she wanted to cry out that she could not do it. But she bit her tongue and sat in the darkness trying hard to pretend she was in a cave on a Girl Scout outing. A quarter of an hour later, to her immense relief, she was released.

Another exercise Christa was not looking forward to was being strapped into a large, round, multiaxis contraption that would spin her in every imaginable direction. Since childhood she had suffered very badly from motion sickness, but she knew it was something she would have to conquer – especially as the candidates would soon be sent up on a dizzying ride aboard a NASA aircraft where they would experience short bursts of weightlessness. With great apprehension she allowed herself to be strapped in and was soon spinning end over end, rotating around all axes, first one way, then the next. When it finally stopped and Christa opened her eyes the room seemed to be spinning violently. She was feeling extremely nauseous and kept stumbling out of control. At that very moment she knew she had lost any chance of being selected.

A further disappointment came a little later, when Christa was asked to give another set of blood and urine samples. Feeling that this was now the end of her hopes of being chosen, she became even more downhearted but tried her hardest not to let anyone notice.

The tests continued at a cracking pace. All of the teachers participated in a high-altitude chamber flight that subjected them to test conditions at the equivalent of thirty-five thousand feet and their reactions to experiments in decompression and hypoxia (a lack of oxygen to the brain) were carefully noted. This in turn led to the daunting flight on NASA's KC-135, a converted 707 aircraft. It was the test Christa dreaded the most, and she felt sure she would become so ill she would be immediately marked down as an unsuitable candidate for space flight.

In an attempt to quell her rising fears, Christa concentrated on an exciting major event the following day. After their flight the teachers had been invited to stand behind some Johnson Center mission controllers monitoring the launch of shuttle *Challenger* on the STS 51-B/ Spacelab 3 mission. She would later say that it didn't matter how sick she got on the KC-135, she wanted to watch the televised launch even if they had to prop her up against a wall.

Because of the stomach-churning aerobatics the pilots routinely performed, the KC-135 had become commonly known as the "Vomit Comet" – and not without reason. A typical flight pattern simulates to a limited degree the experience of being in space. To achieve this it travels to around thirty-four-thousand feet, then dives sharply at a forty-five-degree angle and plunges downward for about forty seconds. During the period when the aircraft is coasting at the bottom of this dive the pilot pulls up the nose of the KC-135 and a weightless condition is produced inside the fuselage. In that half-minute the occupants are as weightless as they would be in space.

Astronauts Dale Gardner and Judy Resnik took the excited but apprehensive teachers on their flight aboard the agency's training aircraft. On the first run Gardner helped the teachers to relax by emptying out a large bag of toys: Frisbees, tennis balls, paper airplanes, and pieces of string. During each period of weightlessness the teachers were able to propel these items around while learning how to move about by pushing against the fuselage walls. Altogether they made twenty-seven climb-and-dive maneuvers and learned a great deal about the effects of weightlessness and how to cope in a weightless environment.

Once she had adapted to the routine Christa actually began to enjoy the experience, spinning around, bouncing effortlessly off walls, tossing paper airplanes, and forming circles by holding hands with the other teachers. But after two hours of this even the strongest of stomachs begins to rebel, and Christa was feeling ill. She was not the only one now suffering, and even former Air Force pilot Mike Metcalf had thrown up.

Christa had to sit down. Her stomach was churning and her face had gone pale. Hot tears of frustration burned her eyes. Kathy Beres, noticing Christa's distress, came and sat beside her and held her hands as the aircraft finally headed back for a landing. Christa could handle any psychological test they placed before her, could show determination in any task she was set, but the onset of nausea was something she could not control.

Despite their queasiness toward the end of the giddying flight most of the teachers were actually disappointed that it was over – and all of them wanted to do more!

Christa would later find out that feelings of nausea toward the end of the two hours on the "Vomit Comet" were quite normal for anyone taking the ride, and her illness did not go against her. She would also learn that the extra bodily samples she had given were only the result of a lab mishandling of the first set of samples, and there were no significant problems when the results were compiled.

Following a shower and rest and a little lunch the candidates were introduced to the waiting press for a media session, and then they were ushered into the Mission Control blockhouse to monitor the launch of *Challenger* at the Kennedy Space Center in Florida. As the launch time approached their excitement grew. All of them wondered how the astronauts on board were feeling as the moment of liftoff approached. Unhappily the launch was shut down at T-3 seconds due to a faulty main engine valve, and the mission was postponed. But, as Christa would later state, it helped and reassured her to understand that a shuttle launch could be safely shut down if there was any danger to the astronauts. Astronaut Dick Scobee, who would command the Teacher in Space mission, later briefed the teachers on the inherent dangers of space flight and answered questions they raised.

After this week of intense evaluation the finalists were sent on a short visit to the Marshall Space Flight Center in Alabama before returning to Washington. Here they were able to view up close the type of massive boosters that would send one of them into space. At the center's Rocket Park they went on a couple of rides, and once again Christa became ill after sitting through a multiaxis spaceball ride. However, she was determined to press on and rejoined the others. Their next challenge was the Lunar Odyssey centrifuge ride, in which the participants would buckle themselves into a seat and be spun around at high speed. But the teachers' laughs were soon quelled when tragedy struck.

As the centrifuge began to whirl a young high school graduate named Greg Walker, not buckled in, tried to move about while the ride was in progress. He was thrown around and eventually slammed through a wall of the chamber near Christa's seat, into the ride's machinery below. Following the teachers' frantic shouts the centrifuge was shut down. To her horror Christa was the one who first found the boy's smashed body and pointed him out to rescuers. Several hours later she was told that the youth had died. That tragic event would haunt her until the day she died.

The next and last step for the flight candidates would be their final interviews. For three days, beginning 15 July, a NASA evaluation committee made up of senior space agency officials interviewed all ten teachers. This committee, under the supervision of NASA executive Ann Bradley, was already armed with the results of the interviews and tests to date. They set out to cross-examine each of the teachers and to come up with the name of the person they felt could best handle the flight, the media, and the resultant fame and who would project the type of wholesome image NASA required of its teacher.

It was one of the worst times for Christa. Not only did she have to cope with the recent tragedy and keep up a brave face in public, but now she also began to worry about how her training and flight would affect her family. Steve had assured her he could cope with being a single parent for a while, but there were other anxieties to deal with. She was going to miss her family enormously, especially little Caroline. She would not be there to comfort Caroline if she woke up crying and would not be there to read to her at night. At a crucial time, when her little daughter needed her to be there, Christa would be off training in Houston and then climbing aboard a spacecraft with the very real possibility that she might never return.

Many nights Christa found it hard to sleep, tossing and turning.

Was she doing the right thing or just being purely selfish in wanting to fly into space? She wasn't an astronaut, for goodness sake – she was a schoolteacher and a mother. Why on earth was she leaving her loving family behind her for those long, lonely months? How would they manage without her? And how would she cope herself with the agony and loneliness of not seeing them for such a long time? What would happen if there was an accident and she never came home again? Her mind in turmoil, she nevertheless tried to carry on as normal.

At the end of the three days the judging panel met to consider their choice. Ann Bradley would later say that Christa had impressed all of them with her calmness and enthusiasm. She was also considered a good "team player" and stood out as an excellent communicator. "All ten teachers were outstanding people," Bradley said, "but some of them concentrated too much on how wonderful it would be to fly in the space shuttle, rather than on how they would use the experience to get teachers excited about the space program. Christa was the one who most clearly understood what we had in mind."

The judges' selection was unanimous. Christa McAuliffe would be America's first teacher in space.

They did not know the result of the judging committee, and so all ten finalists were apprehensive as they gathered just after midday in room 7002 at NASA Headquarters in Washington DC. It was 19 July, and the teachers knew that they would soon be given the name of the successful finalist in advance of the official announcement. They had been told the decision would be made known to them prior to leaving for the nearby White House.

Lunch was delivered but, not surprisingly, few had brought along much of an appetite. They had all been told to prepare an acceptance speech, but Christa was too excited and busy phoning her family and friends. Ann Bradley was there too, and she was waiting for the right moment to break the news to the successful applicant.

Soon after lunch Niki Wenger and Christa were enjoying a quiet chat about cooking, with Christa laughingly describing the way Steve practically lived on corn flakes anytime she was away from home. Ann Bradley had been listening in, and she politely broke into the conversation, looking Christa in the eye. "You'd better tell your husband to stock up on a lot more corn flakes," she said.

Christa didn't quite understand for a moment, and then she gasped. The other teachers were looking on as Bradley smiled. "You're the one," she said. "You're the one going up in space!"

There were a few moments of silence as the other finalists realized their particular dream was at an end, then they rushed over and began to hug and kiss their stunned colleague, who could hardly believe what was happening.

The news became official at the White House ceremony, where it was also announced that Barbara Morgan had been selected as Christa's backup candidate. The two teachers would report to the Johnson Space Center in Houston that September for 114 hours of training over four months.

When runner-up Barbara Morgan was asked if she was disappointed at not being selected as the prime candidate she responded, "I think we're all going to get a chance to fly someday."

Kathy Beres was philosophical about missing out, knowing that at least she and the other finalists would be working for NASA over the next year and had been invited to Cape Canaveral in Florida for the launch. They planned to reunite in August to help write lessons that Christa could present from space and that her teaching colleagues across America could use. "None of us felt we'd lost or were second best or were slighted," Beres said. "There has to be a bit of disappointment, but it's a new beginning. I look forward to going back to the classroom and telling my students about it."

Following the tearful announcement at the White House press conference Christa was ushered outside for the first of her interview sessions. She looked sublimely dazzling in a yellow blazer with a red rose pinned to her left lapel. In answer to one of dozens of question from the throng of reporters, Christa said she was just an ordinary person given a chance to do something quite extraordinary. "I would like to humanize the Space Age by giving a perspective from a non-astronaut, because I think the students will look at that and say, 'This is an ordinary person.' This ordinary person is contributing to history, and if they can make that connection then they're going to get excited about history, and they're going to get excited about the future. They're going to get excited about space."

From the lawn of the White House she was bundled into a cab and driven all over Washington to attend a seemingly endless number of television and media interviews. She was discussing what she would do during the mission, and yet Christa was still coming to terms with the fact that she had been chosen. Following this flurry of engagements she returned to Concord to make arrangements for her lengthy absence.

Coincidentally, Christa's selection as the nation's first teacher in space came during Concord's Old Fashioned Bargain Days. During these three festive days Main Street becomes a pedestrian mall where an array of goods are sold from colorful sidewalk stalls. The NASA announcement lent a special air to the festivities, and the chief organizer asked Christa if she would participate as honorary chairperson. She was thrilled to accept.

The following day Christa was seated in the back of a Mercedes convertible, perched on top of the back seat with her children on either side while Steve rode in the car in front so he could photograph the happy scene. She was wearing a cobalt-blue NASA jumpsuit from Alabama's Space Camp and a red, white, and blue baseball cap with a space shuttle emblem sewn onto the front. The crowd went wild with enthusiasm, applauding, calling out congratulations and good wishes, asking for her autograph, and presenting little bouquets of flowers. Any thoughts she may have had of embarrassment disappeared in this spontaneous display of support and encouragement.

There was a lot to organize in Concord. Aware that she would have to set things right with the students in her classes before she traveled to Houston, Christa carefully briefed her temporary replacement, a good friend named Eileen O'Hara. Fortunately Eileen was familiar with Christa's course The American Woman and shared her enthusiasm for the subject. Christa also arranged to take temporary leave from her charitable and sporting interests and found somebody to take over her Sunday school classes.

Saying goodbye to Steve and the children was the hardest thing of all, and Christa knew she would miss them deeply. Scott and Caroline listened intently as their mother explained where she was going, and why, and how their father would look after them while she was away. Scott was of two minds; while he was thrilled about his mother going into space and becoming famous, he didn't want her to leave them. His birthday was coming up in four days, and like all boys of that age he thought it was important for his mother to be there.

Then little Caroline weighed in, hugging her mother and pleading with her to stay. It must have been the most heart-wrenching of farewells for Christa, and once again she must have felt she was somehow neglecting her family in a selfish pursuit of a dream. Tears flowed as she hugged and kissed all of her family, and a deep sadness set in as she handed Caroline to Steve.

Later, when a newspaper reporter asked five-year-old Caroline how she felt about her mother being an astronaut, she responded with a pout: "I don't want her to go into space, because I just want her to stay around my house!"

September came around quickly, and it was time to begin training for the mission. Christa visited Concord High School on 6 September to say a temporary goodbye to the staff and students and then left Concord for the Johnson Space Center. Barbara Morgan also journeyed to Houston to undergo parallel training, knowing she had to be ready to step in and replace Christa at any stage of the training if an unexpected illness or injury occurred.

Uncertainty set in momentarily as Christa's gaze swept over the dull scenery beyond the space center. It seemed so flat and lifeless after the old-world charm of New Hampshire, but she knew that the surroundings were just an illusion; gathered in this otherwise uninspiring place was the cream of America's space scientists and engineers, and she sensed an energy unlike anything she had ever experienced.

She already missed her home and family but in her usual way was determined to stick it out and do the best she could. At the end of her training was the promise of a fantastic journey humans have yearned to make since the dawn of time, and Christa knew that in a few months she would return to Concord and resume her role as a mother and teacher. This would see her through some very intense months of preparation.

Accommodation was provided in a small, furnished, second-floor apartment in Peachtree Lane, off NASA Road 1 in Webster. Happily, Barbara Morgan was assigned to the apartment next door while payload specialist Greg Jarvis had an apartment in the same building.

Although the media was still vitally interested in Christa's educational flight, NASA restricted interviews to just two hours a week, allowing her to concentrate on the training program. This training, the same as that developed for payload specialist candidates, included numerous activities designed to familiarize trainees with the shuttle systems.

All six crewmembers who would join Christa aboard *Challenger* were highly skilled people in their own fields. The commander was Francis R. Scobee, known to everyone as Dick. At age forty-six Scobee was making his second shuttle flight. Born in the rural town of Cle Elum in Washington State on 19 May 1939, he enlisted in the Air Force as a mechanic at the age of eighteen and by studying hard managed to win a degree in aerospace engineering at the University of Arizona in 1965. This qualified him to become a pilot and an officer. A masterful aviator, he soon began to rise through the ranks. He logged more than sixty-five hundred hours of flight time in forty-five different types of aircraft, including the experimental x-24B, a wingless "lifting body" aircraft that was a forerunner to the space shuttle. During the Vietnam War he flew numerous combat missions aboard a Caribou c-7. Prior to joining the astronaut corps, Dick Scobee was one of the pilots who flew the converted Boeing 747 aircraft that "piggybacked" shuttle *Enterprise* between the landing and launch sites. Selected as an astronaut in January 1978, he had flown as pilot aboard *Challenger*'s flight 41-c in 1984.

Navy Commander Michael J. Smith, making his first flight into space, was also a highly experienced pilot. Born on 30 April 1945 in Beaufort, North Carolina, he fell in love with flying at an early age, watching aircraft take off and land at an airport across the road from his home. He sold chickens and eggs to pay for flying lessons and soloed at the age of sixteen. Smith won an appointment to the U.S. Naval Academy at Annapolis, graduated in 1967, and flew 225 missions during the Vietnam War, flying a-6 Intruder aircraft from the deck of the aircraft carrier uss *Kittyhawk* and earning several medals. He later became a test pilot instructor and was selected by nasa to join the astronaut corps in 1980.

Mission Specialist Judith A. Resnik, who held a Ph.D. in electrical engineering, had previously flown into space in 1984 aboard shuttle *Discovery* on mission 39-a. The thirty-six-year-old Resnik had earlier become part of America's spaceflight history as one of the first six women selected by nasa as astronaut trainees. She was the second American woman and the first Jewish astronaut to achieve space flight. She was born on 5 April 1949 and grew up in Akron, Ohio. She developed a keen interest in math and science, which led to her receiving a doctorate in electrical engineering in 1978 from the University of Maryland. Prior to her 1978 selection by NASA she worked at the Xerox Corporation. Judy and her colleagues were chosen from eight thousand applicants. On her first space flight she was responsible for operating the spacecraft's Canadian-built remote-control (RMS) arm and performed experiments with a thirty-four-meter solar sail unit.

Another *Challenger* veteran was Mission Specialist Ronald E. McNair, who launched two communications satellites from the spacecraft's cargo bay during his 1983 mission on STS 41-B. During this mission crewmember Bruce McCandless became the first human satellite when he performed an untethered "spacewalk" away from *Challenger*. McNair, a judo champion and karate expert who held doctorates in physics and science and was an accomplished saxophone player, became NASA's second black astronaut in space. Born in Lake City, South Carolina, on 21 October 1950, he grew up fascinated by the concept of space travel. He was working at Hughes Research Laboratories in 1977 when he heard NASA was seeking scientists to train as mission specialists. He applied and the following year was selected as part of NASA's eighth astronaut group.

The third mission specialist selected for 51-L was Hawaiian-born Ellison S. Onizuka, a lieutenant colonel in the U.S. Air Force. Born on 24 June 1946, he was raised on the big island of Hawaii in the Kona Coast town of Keopu, a small farming community known for its macadamia nut groves and fields of coffee plants. The world's most active volcano, Kilauea Iki, is a near neighbor. Of Japanese ancestry and the first son of Kona Coast grocers Masamitsu and Mitsue Onizuka, young Ellison loved nothing better than to run barefoot through the coffee fields, play baseball, attend Boy Scouts – he became an Eagle Scout – and fly paper airplanes. Even then he said his future was in the skies. As there were no universities with a program in aeronautical engineering in Hawaii, he later moved to the mainland and enrolled at the University of Colorado. After graduating in 1969 and marrying Lorna Yoshida, he spent eight years as a test pilot and flight engineer with the Air Force. Eventually he applied for the astronaut corps, was accepted in 1978, and flew aboard shuttle *Discovery* on mission STS 51-C in January 1985.

The sixth member of the crew was Gregory B. Jarvis, a balding and athletic payload specialist from the Hughes Aircraft Company. Jarvis had twice been bumped from shuttle missions to make way for politicians who had secured flights into space. His major task aboard *Challenger* was to conduct a series of experiments that would determine the effects of weightlessness on fluid contained in tanks, aiding in the design of future communications satellites. Born in Detroit, Michigan, on 24 August 1944, he received a bachelor's degree in electrical engineering from the State University of New York in Buffalo in 1967 and a master's degree in engineering from Northeastern University in Boston in 1968. He was commissioned as an officer in the Air Force in 1969 and became a specialist in communications satellites. In 1973 he left the Air Force and joined Hughes. A classical guitarist, cross-country skier, and runner, he was openly elated when chosen as a payload specialist to accompany a Hughes satellite into space aboard a space shuttle.

From September 1985 until the winter break the crew trained with an intensity that came from wanting to perform the best possible job. In these early days of mission training Christa harbored an anxiety she knew she had to overcome. She wanted her crewmates to accept her as an integral part of the crew and not as a mere passenger on their flight. It was important to her that they knew her participation in the mission would have a lasting educational impact and was not just a politically motivated joyride. Despite the fact that she would dress and train as an astronaut she wanted to prove she was first and foremost a teacher.

Perhaps her greatest concern was her acceptance by the mission commander, Dick Scobee. He had obviously been aware of the tremendous hype associated with launching a teacher into space, and she already knew how much the astronauts privately resented the thought of any "space tourist" claiming a valuable seat on their spacecraft. She dreaded the very real prospect of spending several months training with Dick and his crew, knowing that behind her back they secretly wished she had stayed in her classroom and left the spaceflight stuff to them.

Happily Dick Scobee was one of the first to reassure Christa that she was a welcome and valued addition to the crew. "No matter what happens on this mission," he told her, "it's going to be known as the teacher mission. We feel that's good, because people will remember what we do." His words of encouragement meant a lot to her.

At first, in preparing for their mission, the crewmembers spent most of their time undergoing individual training. During this period they came to know each other better, and during breaks from the rigors of training they socialized over dinners and went on picnics. It was all part of their bonding as a team – a very important aspect of any space mission, when a group of men and women have to spend several days together in the cramped confines of a spacecraft.

I want to demystify
NASA and spaceflight.

Civilians in Space

5

The genesis of the Teacher in Space program was a NASA initiative to send private citizens into space aboard the nation's shuttle orbiters. It is worthwhile to reflect on the often controversial history of this undertaking.

It was certainly true that the idea of flying private citizens on America's space shuttles did not sit very well with many people within NASA. This was especially true for the astronauts, who not only trained over many years for their missions but also were also fully aware of the dangers associated with spaceflight. They believed they had earned the right to fly the missions and hated seeing valuable seats on any mission given over to someone who they believed had a very poor reason to take the place of a qualified astronaut.

The Soviet Union had already set a dangerous precedent for sending unqualified people into space. The very first was Valentina Tereshkova, a simple factory worker who became the first women to fly into space when she flew as sole pilot aboard the *Vostok-6* spacecraft in June 1963. Although she had undertaken parachuting courses, Tereshkova was vastly underqualified for such a risky venture. She had never even flown an aircraft. But Soviet space bosses and the politicians of the day wanted to send a woman into space to upstage the American space effort.

Although Tereshkova bravely completed her historic space mission it is said she was quite ill during most of the flight. However, the sensational worldwide publicity surrounding her flight severely embarrassed the Americans. NASA had no plans at that time to fly any women into space and, following the successful *Vostok-6* mission, received some of its worst ever publicity because of this strict policy. NASA's administrators were firm in stating that only military test pilots would fly on their spacecraft, which excluded women. In those days women were not permitted to fly in the armed forces of their country.

Despite this policy a group of women pilots secretly underwent NASA-approved astronaut training in a New Mexico laboratory, but the program was eventually cancelled. They had actually demonstrated that in many ways they could handle a spacecraft better than their male counterparts. The women were calmer in emergency situations, they handled the solitude of simulated spaceflight better, and some even had flight and educational qualifications superior to members of the original Mercury program astronauts. However NASA's administrators were not prepared to risk sending a women into space, knowing that if one died during a space mission, the public outcry might even shut down the nation's space program.

A year after Valentina Tereshkova's mission the Soviet *Voskhod-1* flew into orbit carrying three men, the largest space crew of that time. One of the cosmonauts was Konstantin Feoktistov, a civilian scientist who had helped design the earlier Vostok spacecraft. Another civilian carried on the flight was Dr. Boris Yegorov, who specialized in aviation medicine. Three years earlier he had examined Yuri Gagarin after his pioneering space flight. Both Feoktistov and Yegorov had trained for only six months before making their first (and only) flight aboard *Voskhod-1*. The third crewmember was cosmonaut pilot Vladimir Komarov. He would later die during a test flight of the Soyuz space capsule when *Soyuz-1* crashed back to Earth following guidance system problems and a fatal twisting of the spacecraft's parachutes in April 1967.

Hailed as a great engineering feat at the time, the *Voskhod-1* mission was actually the most dangerous ever undertaken – even to this day – and should never have been attempted. In order to cram the three men aboard the stripped-down spacecraft they could not wear protective space suits and there were no ejection seats. Had there been an explosion or fire during the launch sequence the three men would have been unable to leave their spacecraft and would have died a terrible death. During their flight they had very little room to move around in and carried out only some very minor experiments. Thankfully for the three crewmembers the flight lasted just over a day before the deorbiting rockets were activated and their spacecraft returned to Earth as scheduled.

As with the flight of Valentina Tereshkova this flight was nothing more than a perilous propaganda stunt, a sensational attempt to show that the Soviet space program was well ahead of America's efforts. For several years this propaganda tactic worked, with the world in general believing the Soviet space program to be far superior to that of the United States. While America and NASA were continually frustrated by the Soviets' demoralizing spaceflight "firsts," the space agency would never have attempted the many dangerous and totally unnecessary feats carried out in the name of publicity and propaganda.

For years to come the Soviets would continue their policy of sending partially trained, politically correct "guest" cosmonauts into space through their Soyuz program. They selected pilots from Soviet-bloc countries and flew them on weeklong missions to link up with a series of Salyut space stations. The first such "guest" cosmonaut was Captain Vladimir Remek from the Czechoslovakian Air Force, who flew aboard *Soyuz-28* with cosmonaut Alexei Gubarev in 1978. Remek would laughingly refer to himself as the "red-handed spaceman" in later years. It was said that if Remek went to touch any instrument Gubarev had instructions to slap him on the back of the hand and tell him to leave it alone!

In 1983, following the first successful operational flights of America's space shuttle, an advisory group reported to NASA administrator James Beggs on the possibility of carrying private citizens on shuttle missions. Beggs was convinced that it would provide the kind of positive publicity the space program needed at that time. He also wanted to demonstrate that shuttle flights would soon become an ordinary activity.

The report concluded that it would be possible to fly civilians by the mid-1980s, when seats would become available and private citizens with a few months' specialized training could be carried on board without risking crew safety or the mission. The report suggested that flying observers on shuttle missions would increase the public's understanding of spaceflight.

Beggs knew that once the report was released NASA would be

swamped by applications from every corner of the United States. He therefore outlined certain broad requirements. The applicants must be motivated to fly on the shuttle and give strong evidence as to how they would use this unique opportunity. They had to be physically and mentally fit and able to perform one hundred hours of training. As well, any applicants had to be able to adapt to the living conditions and working relationships aboard a shuttle, and they were not to make any later profit from their experience.

An important recommendation of the report was that the first civilian space travelers should include trained observers, who could provide written and verbal accounts of their flights that could be communicated to the public, especially for use in classroom teaching. The decision as to who would fly rested with Beggs, and as he said in 1983, it would not be an easy question.

You have a potpourri of people who want to fly, and you must decide who can go. The first should probably be a journalist or someone who writes for a large-circulation publication. A moviemaker making a documentary might be useful, I think. Whoever goes, our expectations should be modest. The first few years are unlikely to produce something terribly spectacular. The things that come to my mind when the astronauts come back is that the pictures don't do justice to the scenery, the spectacular view, the wonder of looking out, the blackness of space, the black background and the stars, all the soul-stirring kinds of things. Astronauts are by their nature externalizers and basically trained to do engineering and science. They don't have the inclination to look at things with an inner eye. There's a real experience there for someone who thinks in artistic terms to gain real meaning – ideally someone with a bit of the poet in him.

As with all U.S. manned space flights, a special mission emblem was devised for mission STS 51-L (*upper right*). Shuttle *Challenger* is shown with open payload-bay doors, displaying representations of the experiments carried, the TDRS-B satellite, and secondary payloads. The backdrop of the American flag displays seven stars, for the seven crew members, and the shuttle's launch trail emanates from the Kennedy Space Center in Florida. One on-board experiment, Spartan Halley, is represented by the depiction of Halley's Comet, while the names of the five NASA astronauts are shown around the primary circle of the badge. The add-on border at the foot of the badge contains the names of the two payload specialists who were later assigned to the crew, while the red apple is obviously in homage to Teacher in Space Christa McAuliffe. The Teacher in Space Project logo (*lower left*) was designed for use by the ten program finalists.

The spectacular sight that would have greeted Christa on the morning of the launch: space shuttle *Challenger*, floodlit on the launch pad.

Left: Ice and frost coated Pad 39B on launch day. *Facing page: Challenger's* left wing, solid rocket booster, and external fuel tank are seen behind the icy structure.

Two years old. By this time everyone knew the pretty little blond-haired girl as Christa rather than by her given name of Sharon. "Somehow," said her mother, Grace, "it really suited her better." (Courtesy Corrigan family)

A delightful mother and daughter photo. (Courtesy Corrigan family)

The ten finalists in the NASA Teacher in Space Project. *Left to right:* Barbara Morgan, Richard Methia, Kathleen Anne Beres, Robert Foerster, Niki Wenger, Michael Metcalf, Peggy Lathlaen, David Marquart, Christa McAuliffe, and Judith Garcia.

The ten finalists for the Teacher in Space Project pose for a group photo with the KC-135 "zero-gravity" aircraft at Ellington Field, near the Johnson Space Center. Shown are (*standing, left to right*) Richard Methia, Peggy Lathlaen, Barbara Morgan, Christa McAuliffe, Kathleen Ann Beres, Niki Wenger, Robert Foerster, Judith Garcia, and (*kneeling*) Michael Metcalf and David Marquart.

NASA's first six women astronauts pose with a mockup of the Personal Rescue Enclosure (PRE), or "rescue ball," similar to the one in which Christa was later tested for claustrophobia. From left: Rhea Seddon, Kathy Sullivan, Judith (Judy) Resnik, Sally Ride, Anna Fisher, and Shannon Lucid. Behind them is a mannequin attired in an Extravehicular Activity (EVA) suit.

Christa McAuliffe accepts her selection as America's Teacher in Space at the White House, 19 July 1985. To the left of Vice President George Bush is William Bennett, Director of the Department of Education.

NASA's KC-135 "zero gravity" aircraft climbs and dives repeatedly on a parabolic pattern to simulate weightlessness for the occupants.

Barbara and Christa enjoy the experience of weightlessness aboard NASA's "Vomit Comet," a KC-135, while

Weightless training aboard NASA's KC-135 aircraft.

Christa and Barbara relax aboard the KC-135 aircraft and chat with Congressman Bill Nelson, who would fly on the mission prior to the one destined to carry a teacher into space.

Kennedy Space Center, 30 October 1985. Barbara Morgan (*left*) and Christa witness the launch of shuttle *Challenger* on mission 61-A. This would be the last

The day she left Concord High to begin mission training, Christa (wearing her lucky yellow jacket) was given a rousing send-off at the school. (Courtesy of the *Concord Monitor*)

The two teachers in the shuttle training area at Houston's Johnson Space Center.

The crew of the shuttle mission STS 51-L. *Front row, left to right:* Michael J. Smith, Francis R. (Dick) Scobee, and Ronald E. McNair; *Back row:* Ellison S. Onizuka, S. Christa McAuliffe, Gregory Jarvis, and Judith A. Resnik.

A nervous smile and a wave as Christa perpares for her familiarization flight aboard a T-38 with Mission Commander Dick Scobee.

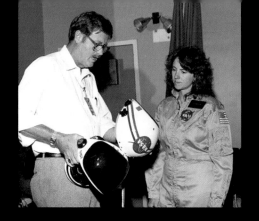

Christa is briefed on the launch/entry helmet by Johnson Space Center's crew systems technician Alan Rochford.

Christa focuses on a test subject as she learns to use the Arriflex motion-picture camera carried on shuttle flights.

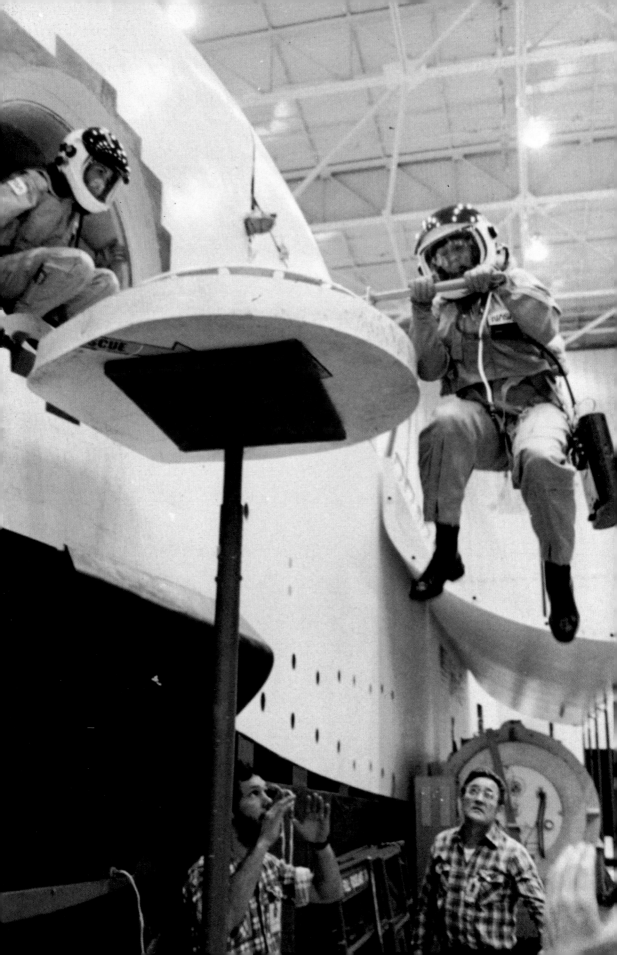

Christa swings from a hatch extension during emergency escape training as Judy Resnik awaits her turn.

Christa poses on the pilot's seat in the shuttle mockup at Johnson Space Center. Note the sign on the rear of the seat that has been altered to read "muck up."

On all shuttle flights the crew has to exercise daily. Here Christa practices walking on a treadmill exerciser carried on board. The astronauts have to be strapped down in order to exert pressure on the treadmill and, more importantly, so they don't float away.

Challenger's crew received slidewire escape training at Pad 39B. If a prelaunch fire or explosion required an immediate evacuation of the launch tower by the astronauts, they would board these steel baskets and slide down cables to the ground. Ron McNair, Greg Jarvis, and Christa occupy the front basket; in one at the rear are Judy Resnik and Ellison Onizuka.

The flight deck of a shuttle mission simulator, with four of the 51-1 crew members seated in their actual takeoff positions. Left is Pilot Michael Smith, behind him is Mission Specialist Ellison Onizuka, seated next to Mission Specialist Judy Resnik. At right is Mission Commander Francis (Dick) Scobee.

On the shuttle simulator's mid-deck, back-up Barbara Morgan reads a procedures list while Payload Specialist Christa McAuliffe and Greg Jarvis sit in their takeoff positions and Mission Specialist Ron McNair peers out

Launch day, and Mission Commander Dick Scobee leads his crew out to the transfer bus that will carry them to the launch pad. Judy Resnik is next, followed by Ron McNair, Mike Smith, Christa McAuliffe, Ellison Onizuka, and Greg Jarvis.

Challenger's solid rocket boosters (SRB) ignite, and the spacecraft is committed to launch. Note an ominous puff of black smoke at the foot of the right-hand SRB; this is where the burn-through occurred that eventually caused the explosion seventy-three minutes into the flight.

Liftoff, and space shuttle *Challenger* clears the launch tower.

Fifty-nine seconds into the flight, and a plume of flame becomes evident on the right-hand SRB.

Seconds later the plume of flame has intensified dramatically.

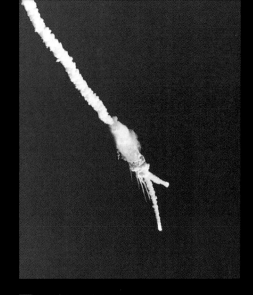

The shuttle's main tank explodes with devastating force, and the two SRB's are propelled away from the billowing, expanding cloud of smoke that indicates the area of the explosion, while debris showers back to earth.

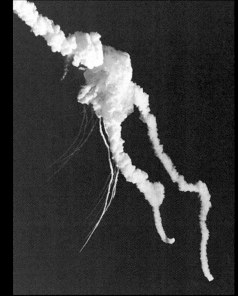

S. Christa McAuliffe and
Barbara Morgan.

Certain operational restrictions were placed on any civilians chosen to fly into space. They would not be permitted on the first flight of a new shuttle, they could only go on a flight where their presence would be of some benefit to the American people, and they could not occupy seats on flights carrying highly classified military payloads.

As expected, the applications poured in. They came from famed explorers, world leaders, noted film and television actors, and members of Congress, and even the name of former President Jimmy Carter was put forward as someone willing to fly. Some of the famous personalities known to have applied are explorer Jacques Cousteau, singer and songwriter John Denver, and comedian Bob Hope.

Deciding who was going to be the first private citizen to fly into space was proving to be a huge administrative nightmare, but the process continued. Eventually, in September 1984, the decision was actually made by President Ronald Reagan. He informed NASA that the first civilian into space should be a teacher. High on the agenda of his administration was an emphasis on education, especially technical education, and he was also a firm supporter of the space program. It was a solid, logical choice, and Beggs readily agreed. The process of finding the nation's first teacher in space began.

For the most part Americans were delighted by President Reagan's commitment to send a teacher into space, but there were grumblings from dissatisfied journalists. They had fully expected that one of their own would become the first spaceflight observer.

Interestingly, a private citizen outside of this selection process would make the first of his three eventual spaceflights in August 1984. Charlie Walker was an engineer employed by the McDonnell Douglas Corporation, who had developed a machine designed to manufacture certain pharmaceutical drugs in space. The company paid NASA the cost of training Walker, and when he was launched aboard shuttle mission 41-D he became the first commercial space traveler. He would make two more space flights for McDonnell Douglas, in April and November of 1985.

A second commercial space traveler, Bob Cenker, was added to the crew of mission 61-C in January 1986. He was a satellite engineer with RCA's Astro-Electronics Division. During his one and only space flight he assisted in the deployment of RCA's Satcom Ku-1 satellite.

As the selection process to find the first teacher in space continued, NASA once again found itself in the center of controversy. In November 1984 Beggs invited a politician to fly on an upcoming space shuttle mission, and there were strong allegations that NASA was trying to gain political favors by giving away seats on the shuttle.

The politician in question was Utah senator Jake Garn, a former navy pilot and chairman of the Senate committee that oversaw NASA's budget. Beggs argued that NASA had always planned to offer space-flight opportunities to VIPs and ordinary citizens and Garn was not only an appropriate choice but he also met all the qualifications. Under a storm of protest Beggs finally agreed that Garn would be the only politician to fly into space in the foreseeable future.

Prior to Garn's flight a civilian employed by the U.S. Navy flew as an oceanographic observer aboard shuttle *Challenger* on mission 41-G. This payload specialist was Australian-born scientist Dr. Paul Scully-Power. During his eight-day flight he studied and photographed the oceans of the world and also carried out classified visual experiments for the navy. As with the flights of Walker and Cenker, there was at least some merit attached to Scully-Power's journey into space. He made several important discoveries about the influence ocean currents have on global weather, leading to better means of forecasting extremes such as flooding, typhoons, and drought.

Jake Garn eventually flew on mission 51-D in April 1985. Due to mounting delays in the shuttle program and subsequent crew and payload changes he bumped from the flight a civilian payload specialist who would later be reassigned. This man was Greg Jarvis from the Hughes Aircraft Company, who had been scheduled to deploy his company's communications satellite from *Discovery*'s cargo bay.

Senator Garn spent seven days aboard shuttle *Discovery* and participated in several low-key medical experiments. He was ill during much of the flight, however, leading many people and astronauts to wonder at the wisdom of having him on board in the first place.

Despite his earlier assurances that Garn would be the only politician to fly in a shuttle, Beggs drew further criticism in September 1985 when he announced that yet another politician would fly. This time the space passenger was Florida congressman Bill Nelson, another politician who had responsibilities for overseeing NASA's budget. He would fly on mission 61-C just four months after Beggs's announcement, in January 1986. A late inclusion into the crew, he took the place of the luckless Jarvis, who now found himself assigned with the Hughes satellite to yet another flight. Jarvis would eventually find himself as part of mission 51-L, under commander Dick Scobee, on a flight destined to carry the first teacher into space.

To say that the astronauts were becoming increasingly annoyed at the publicity flights being given to politicians and other VIPs would be an understatement. Even a Saudi prince had been given a ride aboard shuttle *Discovery* in June 1985. Every seat on every flight was seen as highly desirable, and it seemed that years of hard work could be swept away by people pulling the right political strings. The contributions these "guests" made to each flight were seen as having very little value, and the astronauts knew things would only get worse as other prominent people jockeyed for seats on upcoming flights.

Despite all the criticisms NASA pressed ahead with their plans to fly observers into space. In October 1985 they announced a program to select the first Journalist in Space, and entries flooded in from across the United States. The selected journalist would fly on mission 61-I in late 1986 under commander Don Williams, whose pilot would have been Mike Smith, on his second scheduled flight.

Reporters, columnists, editors, weathermen, and photojournalists all sent in their applications. The minimal requirements they had to meet were at least five years' news experience, citizenship of the United States, and the ability to hear a whispered voice from three feet. All government employees were banned.

Some of the more prominent applicants were NBC's Tom Brokaw and John Chancellor; CBS's Walter Cronkite; the host of ABC's *Good Morning America*, David Hartmann; fellow ABC commentator Geraldo Rivera; and author Norman Mailer. The print media was represented by such applicants as the *Washington Post*'s Kathy Sawyer and John Wilford Noble of the *New York Times*.

The Association of Schools of Journalism and Mass Communication in Colombia, South Carolina, was appointed by NASA to administer the selection and forward their recommendations to the space agency. Eventually 1,703 applications were received for consideration. Plans called for one hundred regional semifinalists to be selected. By March 1986 this list would be whittled down to forty. By 5 April five finalists would be selected and would then undergo further screening and tests at the Johnson Space Center. The most suitable candidate would then be chosen, and on 17 April NASA would announce the winner's name.

Without doubt one of the sentimental favorites with the American public was CBS special correspondent Walter Cronkite, then sixty-nine years old and one of the most respected and trusted figures in the American media. He knew his age was against him but was determined to give it his best shot, as he told a reporter from *Life* magazine. "I've been in bombing craft over Germany, made landings on hostile beaches. I am calm under duress. And I seem to come out of it without disgracing myself or my colleagues. As captain of a vessel [a forty-two-foot yacht], I balance personalities. Ocean sailing is not dissimilar to the shuttle environment. I just hope they're farsighted enough not to consider age. I want to cover the story!" Cronkite came close to realizing his dream. When the names of the forty finalists were announced his name was still on the list of candidates.

The Department of Defense was also jockeying for seats aboard the space shuttle in order to conduct experiments with military applications. They wanted to send into space meteorologists, geologists, and oceanographers who would gather military intelligence information on highly classified missions.

As the head of all Air Force space programs, Air Force undersecretary Edward (Pete) Aldridge was offered an assignment as an observer–payload specialist on mission 62-A, scheduled for launch in July 1986. This flight was intended as the first shuttle mission to take off from the new launch facility at Vandenberg Air Force Base in California, then under construction.

Singer and songwriter John Denver was still pressing his claim for a seat aboard a shuttle. He said he wanted to write a song about his experiences in orbit. He would later approach Soviet space officials and offer to pay several million dollars for a flight to their Mir space station, but he never succeeded in his dream of flying into space.

Film producer George Lucas, of *Star Wars* fame, was also interested in the prospect of viewing the Earth from space. There were even calls in some quarters for President Reagan to fly aboard a shuttle, but this was never a serious consideration.

Still more names came forward. In November 1985 it was disclosed in the British press that *Pollyanna* actress Hayley Mills had put her name down for a three-day flight on a shuttle "around 1996," as part of NASA's Citizen in Space program. It was reported in the *Star* newspaper that the cost of her flight (at 1985 prices) would be around forty thousand pounds and would "include a program of two months of fitness tests and orientation training. Miss Mills is hoping to become the first actress in space."

The question of flying civilians and, especially, politicians would be raised again many years later, when seventy-seven-year-old senator John Glenn was launched into orbit aboard shuttle *Discovery* on 29 October 1998. For Glenn it was a triumphant return to space, coming thirty-six years after his solitary flight aboard the cramped Mercury capsule *Friendship 7*. Ostensibly Senator Glenn was given his place aboard the shuttle to study the effects of spaceflight and weightlessness on an elderly test subject, and while these tests were certainly performed, most people knew that this flight was one to celebrate the life and achievements of America's most popular astronaut.

This time very few people raised any objections, and in fact America got right behind the flight with more people viewing the launch live from vantage points around the Kennedy Space Center than had been seen since the days of Apollo. Banner headlines in newspapers, rejoicing in the patriotic fervor associated with the event, expressed the nation's encouragement and good wishes.

America has always been proud of John Glenn, and Glenn had always wanted to make another space flight but had been prevented from doing so by President John Kennedy. Kennedy had told NASA after Glenn's historic three-orbit flight that he didn't want to risk the life of America's newest hero on another space shot, and Glenn reluctantly left the space agency to go into private business and then politics. He even had a shot at the presidency but for once fell short in this lofty ambition.

During his nine-day STS-95 mission even Glenn's crew stated their pride in having him on board and were more than happy to see the mission assume the unofficial title of "John Glenn's flight."

Back in the mid-1980s, however, things were far different. There is no doubt at all that NASA's astronauts were becoming increasingly concerned and even irate about the number of seats being offered to virtual "space tourists" on upcoming missions, but they were not permitted to express these views publicly.

It was into this hostile but unspoken environment that Christa McAuliffe found herself as a civilian with a seat aboard a shuttle flight. Little wonder she was deeply concerned about her acceptance by the NASA crew.

*You learn things so that you can under-
stand the shuttle. But it's not like I have
to figure out how it was put together.*

6

For Christa the most difficult part about astronaut training was the separation from her family, although it helped that Dick Scobee, Mike Smith, Ellison Onizuka, and Ron McNair had their own families quartered in nearby communities. She got to know all of them and in particular June Scobee, who was also a teacher. They became good friends.

In addition to her official spaceflight training program, Christa carried out her own exercise regime, running several kilometers most days.

As part of her training she underwent a dizzying flight with Dick Scobee in one of NASA's T-38 jet aircraft while Barbara Morgan was paired in another with pilot Mike Smith. Apart from essential ejection-seat training, it was necessary to place the trainees in the stressful environment of a narrow cockpit while enduring some high-G aerobatics.

The two teachers were flown through maneuvers, which gave them the chance to experience around five G's or the equivalent of five times their own body weight. During a shuttle takeoff, they would undergo

3.5 G's. Christa was told that if these aerobatics or the high-G forces caused her extreme discomfort, then she wasn't ready to go into space. As it turned out her reactions were quite normal under the circumstances, and Dick Scobee did not feel she was overstressed at any time. He and Smith would tell their backseat passenger when they were about to break the sound barrier or do a barrel roll or some other "breaking in" maneuver. The two teachers coped well and were even permitted to take the controls for a time and put the T-38s through some rolls and dives. Once again there were no problems, and both students were elated.

Further parabolic flights aboard NASA's KC-135 aircraft were scheduled, and both were keen to conduct some simple tests during the short periods of weightlessness, as Christa later revealed. "We were at first trying to be very serious. I said to Barbara, 'Let's get acclimatized on the first parabola. Then we can see how our experiments will work in zero G.' But then it was hard to stay serious. Leapfrog? Why not? We figured kids would love to see that. But there were these serious NASA people there. I don't think a lot of them would leapfrog on a KC-135!"

It was not all fun and games; there were some essential things they had to learn during these brief periods of weightlessness, such as how to eat and drink without the aid of gravity. They quickly learned that simple everyday activities have to be done differently when the force of gravity is removed.

Christa flew three times aboard the "Vomit Comet." She and Barbara Morgan were also involved in a packed program of self-education, including long, intense simulations and emergency procedures, even down to such requirements as the operation of a sixteen-millimeter camera that they would be using in space. Dozens of manuals had to be studied and a host of duties learned. Barbara Morgan stated that "Mentally, the training was rigorous but fun and exciting."

Ellison Onizuka, who would film two of Christa's lessons, was pleased to have the bright young teacher on his crew. "She's a very personable person, an asset to the mission," he told a reporter from *usa Today*. "I think this is a big step to getting word to youngsters that one day space flight will be as routine as the 747." Sadly those words would prove to be quite wrong.

Despite the intensity of her training Christa took time out to show her family around the center whenever they could arrange an official visit. In November Steve brought their children to Houston for a joyous family reunion. He had found it difficult balancing his work with home duties and coping with two very active children. Fortunately their neighbors had been of considerable help in looking after Scott and Caroline when needed, giving them meals, and taking them on shopping trips. Christa was delighted when he told her that the experience had actually brought him closer to the children, and though they all missed her he was enjoying the temporary role of a single parent.

Scott and Caroline had been bug-eyed with excitement on their first trip to Houston. A proud mom took them on a tour of the center's facilities, where they saw real and mock-up spacecraft, viewed moon rocks, were introduced to several astronauts, and ate lunch at the center's restaurant. Christa's parents arrived on a later visit, and she told them an amusing story about her daughter's impressions of the space center. Asked what she had enjoyed most about the day,

Caroline had pondered the question before announcing in all seriousness, "The tuna fish sandwich at lunch!" Christa laughed fondly at the memory and said it had really brought her back to earth!

By the time of a training break over the holiday season Christa had worked hard. She and the rest of the crew had been up at eight every morning ready to tackle what was mostly individual training, but as the weeks passed and the workload intensified they began to train together and function as a crew. In the final forty-five days before launch they went through repeated mission simulations and familiarizations. Launch day drew closer.

By now Christa and Greg Jarvis, who was married to his college sweetheart Marcia, had become good friends. Separated from their families during the training period, they supported and encouraged each other, making their otherwise grueling schedule a lot more tolerable. On some evenings when they were not required to study one of the endless number of systems manuals, they would relax and chat over a game of Trivial Pursuit.

Thanksgiving and Christmas breaks meant the crew could spend some time at home with their families, and these occasions were greatly appreciated. There were also times when the crew would find themselves enjoying a short but welcome period of relaxation and days when Christa would surprise them by baking up some delicious apple pies.

By now hundreds of letters had poured into Houston for Christa, expressing delight at her selection and offering good wishes for a safe and enjoyable flight. Many asked for a signed photo, and she happily obliged. As a teacher Christa was used to signing dozens of reports and making assignment notations every week, so responding to this mail did not come as an unwelcome chore. If anything she enjoyed the letters sent by young people. Handwritten replies went to everyone, most similar in content to a letter Christa sent to Simon Vaughan of Ontario, Canada: "I had followed the program when the first satellite was launched, and now looking at the shuttle it doesn't seem possible that so much could have happened in such a short period of time. Of course I had to apply! The chance to be a part of that history was so exciting. I am looking forward to seeing the earth from two hundred miles up – I'm hoping to take many pictures of my new experience. And I'm looking forward to sharing this with everyone when I return."

Meanwhile the eight teacher finalists had convened over summer. As part of their commitment to the educational flight they had planned and drawn up ideas for eight elementary experiments to be conducted during Christa's mission. She also met regularly with NASA's Bob Mayfield and discussed her own series of simple scientific experiments, which she intended to carry out during the live TV transmissions. Later Bob Mayfield would express his total admiration for Christa, saying, "She turned out to be an A-plus student."

The Ultimate Field Trip was the first lesson Christa planned to beam down, on Day Six of her mission. Two lessons were scheduled, which would not only be filmed but also transmitted live from orbit by the Public Broadcasting System to schools equipped with satellite dishes. These lessons, under the project title of "Classroom Earth," had been carefully devised so that students could follow parallel lesson plans provided by NASA and conduct the same experiments on Earth. The only difference was that Christa's lessons would be conducted in microgravity, and the students would try to replicate them at 1G. Participating schools would receive in advance of the mission some educational materials, a television schedule, orbital map, a Shuttle Prediction and Recognition Kit (SPARK), and other information to prepare teachers and students wanting to follow all aspects of the flight. Barbara Morgan, as the back-up candidate, was scheduled to act as moderator for the special broadcasts.

Christa would conduct a number of demonstrations during her flight, all used as part of several educational packages to be prepared and distributed after the mission. She was especially keen to conduct three-dimensional experiments in magnetism – an activity that then could be performed only in a weightless environment (these days it can be carried out in laboratories). Students back on Earth would try to carry out the same experiment and note how gravity can greatly influence the result.

There would also be a demonstration of how everyday procedures can prove to be unexpectedly difficult in space, such as the simple act of using a screwdriver. On Earth a screw can be turned quite easily when pressure is exerted, but in space an astronaut's weightless body has nothing to act on. Instead of turning the screw, the astronaut would actually revolve around it. To overcome such problems they have to use foot restraints positioned throughout the shuttle.

Another planned experiment involved a sealed plastic bag of M&Ms and marshmallows to demonstrate how materials with different densities mix in space. An Alka-Seltzer tablet inserted into some water would show that bubbles just don't rise to the surface of a liquid in space.

As part of the Ultimate Field Trip Christa would take earthbound students on a tour of *Challenger*, introducing them to the crewmembers and explaining how human beings live and work in space.

Through the eyes of the camera Christa would visit the flight deck and point out the orbiter's highly advanced computer system. Then the camera would be pointed out of the rear-facing windows into the vastness of the payload bay, and Christa would describe several prominent features. Although broadcast live these activities would be retained on film, a narrative added by Christa on her return, and the finished package released as an educational aid.

Backup teacher Barbara Morgan was given the task of explaining these and other facets of Christa's televised lessons at a news conference. "From [the flight deck] she will float down, she thinks head-first at this point, to the middeck area where she will show the waste control system, better known as the bathroom. She'll talk about the sleeping bag and how astronauts sleep in space, and she'll talk about the galley and how the cooking is done. She'll show the hatch to the airlock and the treadmill. So, basically, the lesson is showing how we live in space."

Christa's second lesson, titled "Where We've Been, Where We're Going," would examine why humans venture into space, the history of this scientific endeavor, and where we plan to go in the future. Using models of the Wright brothers' aircraft and the space shuttle, she would discuss the past eighty years of human flight activity. There would be a description of spin-offs and benefits that have evolved from the space program, a list of ways in which the modular space station would change the lives of human beings, and mention of the advantages and disadvantages of manufacturing in a microgravity environment.

A hometown newspaper, the *Concord Monitor*, had been following and reporting on Christa's activities, and the editors decided to run a competition among local school children. The winners would be given a chance to pose a space-related question to the nation's favorite teacher. The competition was called "A Student's Guide to the Space Shuttle: Christa's Challenge."

Following is a selection of questions and Christa's responses:

q: Can you blow hard through your lips almost like whistling and propel yourself backwards?

a: I'm not sure. We tried it on the kc-135, but it didn't work. We'll try it again on the shuttle. Greg Jarvis doesn't think you can propel yourself by blowing, but he does think that holding a balloon and then releasing [the air in it] should move you across the deck. We will try that.

q: Have you tried any of the space food?

a: The space food tastes good. We have cookies, dried fruit, pudding, and granola bars that look just like the food you put in your lunch box. The meats and vegetables resemble the frozen foods in the boiling pouches that you can buy right at the store.

q: What is the temperature in space?

a: Very cold and very hot. The temperature ranges from minus 275 degrees [Celsius] in darkness to plus 275 [Celsius] in sunlight.

q: Which experiment do you think will be the most interesting?

a: The magnetism experiment is exciting. There's a big, clear plastic tube with an electromagnet in the middle. When Barbara Morgan and I took the magnet up in the KC-135 to test it, the pattern around the magnet was wonderful.

Q: What are your fears about going on the space shuttle?

A: I truly do not have any fears. I'm excited about the trip and am thrilled to have the opportunity.

Q: Will you share your journal when you get back?

A: My journal will allow me to have good recall of everything that I'll be experiencing. I'm hoping to share my thoughts with as many people as possible when I return.

Q: What one thing do you want to discover up there?

A: Personally, I'd like to see how I do in a close, weightless environment with six other people. I'm not going with any preconceived ideas so that I won't ignore anything that happens.

Q: What would happen if you got sick in space?

A: Since you are in a weightless state you need to use a bag and close it up quickly. Many people do not feel well because there is a fluid shift – you feel bloated and there is no up or down. This nausea is gone in one to three days.

Q: What did you like least about your training?

A: There's nothing that I don't like. Everything is new and fun, and remember that I'm getting ready for the ride of a lifetime!

One of the myriad things Christa had to consider was the personal effects she would carry with her into space. Each flight participant is given permission to take a limited number of items in a small satchel known as a Personal Preference Kit or PPK. Without exception these items have to be declared to NASA and cannot exceed a certain weight. In her PPK Christa decided she would include a favorite Bob Dylan tape, a Girl Scout pin, her sister Betsy's ring, Steve's VMI class ring, her daughter Caroline's gold crucifix and chain, and her son's stuffed frog, named Fleegle. She would also be wearing her grandmother's gold watch.

The year 1985 finally gave way to the new year, and the crew was now in the last stages of training. The public, as ever, were keen to know how Christa felt about her upcoming flight. A reporter from USA Today asked about her feelings now that the flight was imminent. The reply was cautiously enthusiastic.

I'm not sure how I'm going to feel sitting there waiting for take off and those solid rocket boosters ignite underneath me and everything starts to shake. It's kind of like the first time you go on a carnival ride. You've said "I've got enough courage," and you're really excited about doing this and conquering your fears. The training is not physically rigorous. Mentally, it is. You learn things so that you can understand the shuttle, but it's not like I have to figure out how it was put together.

The crew is unbelievable; very professional, very concerned about how I am fitting in with them, trying to make me part of the team. Judy Resnik, for example, has a doctorate in electrical engineering. It's mind-boggling that she knows all those circuits and can figure all of this out. Not only will I be teaching my lessons and doing activities, but I'm also helping; it's going to be fun.

I've talked with my kids about seeing the launch – some astronauts' kids have had bad reactions [but] I think they'll be okay. Caroline is six and Scott is nine. They are a bit more aware of rockets and how they behave. Caroline doesn't like loud noise, so I told her it can be loud and that bothers her.

I was delighted that a teacher was chosen as the first space participant because there are so many of us who have daily contact with people. To think that teachers were finally recognized as the good communicators they are.

I'm hoping that everybody out there who decides to go for it – the journalist in space, the poet in space – whatever the categories, that you push yourself to get the application in. I'm hoping there are going to be more people down the road who are going to apply for it. When you think of the future, there are going to be more people going into space, and they are going to be the kids who are in our classrooms. It's a wonderful thought!

I still can't believe that I am actually to be going into that shuttle. It just really doesn't seem possible. Maybe when I'm on the launch pad it will.

7

Thursday afternoon, 23 January 1986, three gleaming jets swooped low over the Kennedy Space Center flying in close formation. They began their final approach and then one after another touched down smoothly on runway 15. *Challenger's* crew had finally arrived for their mission.

Canopies over the cockpits opened upward as the sleek T-38 jets taxied onto the tarmac, and a waiting crowd of reporters and photographers held their hands over their ears to block out the shrieking noise of the approaching engines. Patiently they watched as the pilots shut down their engines, then six helmeted figures prepared to clamber down from their cockpits.

Dick Scobee and Ellison Onizuka scrambled out of the first jet, then Mike Smith and Judy Resnik from the second. Ron McNair climbed down from the rear seat of the third T-38 while the pilot, Chief Astronaut John Young, spent additional time in the cockpit writing up his logbook. This day belonged to the *Challenger* crew, and he wasn't about to intrude on their arrival ceremony.

Greg Jarvis and Christa had already flown in from Houston aboard one of NASA's Gulfstream executive jets, and all seven crewmembers were soon assembled for the traditional Astronaut Arrival press conference. Dozens of cameras were clicking and whirring as eager photographers captured the happy scene.

Eventually each of the seven stepped up to a battery of microphones in turn and gave a brief speech. All were dressed in their cobalt-blue training suits and wore shiny, black, high-lace boots. All, that is, except for Christa, who was wearing a more comfortable pair of gray jogging shoes. She was ecstatic and clearly bursting to go on the flight. Her joy was easily evident, and she just kept grinning and waving at the photographers.

When she finally stepped up to the microphones the questions came thick and fast. Asked if she was ready to go into space she smiled broadly and replied, "No teacher has ever been better prepared to teach a lesson!"

That night the *Challenger* crew and their families enjoyed a now-traditional prelaunch party at a secluded but closely guarded beach house not too far from the launch site. When this was over the families were taken back to their motels, and the crew returned to their quarters in the Operations and Checkout Building at the cape.

The following afternoon commander Dick Scobee was informed that *Challenger*'s launch was delayed twenty-four hours, until Sunday, due to a fierce dust storm at an emergency landing site near Dakar, the capital of Senegal in western Africa. Under NASA's strict safety rules this meant a postponement of STS 51-L.

Saturday night brought equally bad news, with shower squalls predicted across the cape the following day. For safety reasons space shuttles could not be launched in rainy conditions, so the launch was officially postponed until Monday morning. On Sunday, to everyone's annoyance, the skies remained clear until well beyond the scheduled launch time. The rains did not arrive until the afternoon.

Following this postponement Christa called her parents and told them she was "rarin' to go." Ed and Grace Corrigan, together with eighteen third-grade students from Concord, were at the cape ready for liftoff. Steve was also sitting out the delays with Scott and Caroline.

On Sunday afternoon the astronauts and their spouses enjoyed an early dinner together in the crew quarters. At the end of the meal Ron McNair produced a small surprise – a magnum of champagne with a representation of *Challenger* etched into the glass. He had also brought along the etching pen and this was passed around so that each member of the crew could engrave his or her name on the bottle. The champagne would be opened on their return to Earth.

The crew was more than ready, but the weather was now giving more cause for concern. A cold front, coupled with shower activity, was inching its way down the Florida peninsula. This in itself was not sufficient to prevent a launch, but once the ascending orbiter reached high speed even the slight drizzle could seriously damage the shuttle's thirty-four thousand fragile thermal tiles.

Sleep did not come easily to Christa that early evening, so she got dressed and gently tapped on Greg Jarvis's door. He too was having trouble sleeping, so they decided to take a pleasant sunset stroll around the space center. On leaving their quarters they came across a pair of bicycles they could borrow, and the two friends rode around the center at an easy pace. They were hopeful about the prospect of good conditions the next day, but as twilight descended their attention was on the skies and the conversation often turned to the chill northwesterly breeze. As they headed back both wondered if the launch would take place in the morning.

Monday dawned with an air of high expectancy. The Florida skies were cool but clear, and word from the launch team seemed to indicate everything was go for a launch.

For the second time the crewmembers ate breakfast, took the transfer van out to Pad 39B, and were assisted into their narrow chairs on *Challenger*'s flight- and mid-deck by members of Lockheed's launch support team. Astronaut Sonny Carter was also there to assist as part of this team. As commander, Dick Scobee occupied the left-hand seat on the flight deck with pilot Mike Smith to his right. Behind them sat Judy Resnik and Ellison Onizuka. Below, on the mid-deck, Christa was strapped in tightly with Greg Jarvis to her right and Ron McNair by the crew hatch. With the shuttle pointed skyward for launch they were actually seated horizontally on their chairs – a most uncomfortable position when held for any length of time. However, the crew hardly noticed this discomfort as they ran through their predeparture checks.

Challenger's countdown proceeded smoothly. Their work done, the Lockheed support team crawled out of the orbiter and closed the crew hatch. Then things started to go wrong; a microswitch indicator warned that the hatch's locking mechanism was not correctly seated, although it was believed that the switch itself was at fault. In order to check the indicator an exterior handle had to be removed, but one of the bolts jammed and an urgent order went out for a cordless drill. Meanwhile the countdown was halted. When the drill finally arrived it was found to have a dead battery. To compound the problem there were no replacement batteries, and the bolt had to be cut off using a hacksaw.

Greg Jarvis began to complain that one of his legs was falling asleep, while Christa managed to doze as the niggling problems continued outside their spacecraft.

Unfortunately the winds around the cape began to pick up once again, gusting to fifty kilometers per hour. Now the crosswinds were becoming so strong that they would prevent the crew from being able to perform an emergency return to the landing site at the cape. Under NASA's strict guidelines that was it – the launch was scrubbed. They would try again the next day.

Everyone was disappointed and frustrated at the reason for the delay; canceling a launch over a battery worth just a couple of dollars was a very costly exercise.

On Tuesday morning, 28 January, Dick Scobee took a deep breath as he left the pad elevator and entered the canvas-covered White Room, ready to try again. He looked out at the blue sky. "This is a beautiful day to fly!" he remarked, and one of the crew support team suggested it was a little cold. Scobee smiled and shook his head. "Nah . . . that's good, that's great," he replied.

Johnny Corlew, a member of the support team, presented Christa with a big, polished Delicious apple. She beamed at him and held the apple in front of her face for a moment before handing it back. "Save it for me," she said, "and I'll eat it when I get back."

Soon they were all strapped into the horizontal seats, once again carrying out their communications checks. Judy Resnik could no longer restrain her excitement. "Cowabunga!" she exclaimed over the intercom, and the rest of the crew laughed, all except Ellison Onizuka, who didn't like the cold. "My nose is freezing!" he complained.

The checks continued, and the crew chatted among themselves, quietly confident, although the cold outside was a major topic of conversation. Finally it came time to close up once again, and the heavy hatch door was latched and sealed. This time everything went as scheduled. It was a little before nine o'clock and the seven crewmembers of *Challenger* were ready to go, on the twenty-fifth space shuttle mission.

Dick Scobee and Mike Smith sat at the control panels, smoothly running through last-minute checks as the countdown clock continued to wind down the minutes. They knew that launch was the trickiest time to control any failure in the shuttle's systems, so they kept a close watch for any indication of a malfunction. It would be too late once the twin solid rocket boosters ignited as the countdown reached zero. For the first two minutes and eight seconds of the flight they would be completely at the mercy of technology, catapulted ever faster into the Florida skies atop three million kilograms of thrust.

At four minutes to liftoff, with the outside temperature now sitting at four degrees Celsius, mission control gave the crew a reminder to close the visors on their helmets. Thirty seconds later the shuttle began to operate under its own electric power.

"Ninety seconds and counting," advised the NASA commentator, his voice transmitted over loudspeakers to the hushed crowd in the viewing stands six kilometers from the launch pad. The countdown clock kept running. Then there were just a few seconds remaining.

"10, 9, 8, 7 . . . we have main engine start!"

Below *Challenger* the three main engines burst into life, sending a huge cloud of steam billowing out from beneath the launch pad as the liquid oxygen and liquid hydrogen fuels combusted.

"4, 3, 2, 1 . . ."

As the countdown reached zero the solid aluminum powder fuel in the two solid rocket boosters ignited explosively and white-hot flames shot downward. At the same moment restraining clamps were released explosively, and *Challenger* had begun its voyage.

Unnoticed at the time, a small puff of black smoke burst from a joint toward the bottom of the right-hand booster. Shuttle *Challenger* was doomed to destruction.

I'm not sure how I'm going to feel sitting there waiting for takeoff, and those solid rocket boosters ignite underneath me and everything starts to shake.

"The Craft Has Exploded!" 8

At 11:38 A.M. on the morning of 28 January 1986 the ground surrounding the Kennedy Space Center began to tremble as *Challenger's* solid rocket boosters roared into life. Slowly the winged spacecraft rose into the azure sky atop a brilliant pillar of white-hot flame. Huge clouds of black, orange, and white smoke billowed outward as *Challenger* surged upward, clearing the launch tower. A massive roaring noise echoed across the cape.

Thousands of spectators on the ground cheered as the shuttle thundered away from the launch pad and climbed into the cold, clear Florida sky, arching out over the Atlantic. Atop a NASA building five kilometers from Launch Pad 39B Steve McAuliffe was watching from a special family viewing area, off-limits to press photographers. With him were Scott and Caroline. Christa's parents and other family members were in a separate VIP section, having agreed to sit with Scott's third-grade class so the press could take photographs.

Everyone shouted out proudly as *Challenger* soared skyward trailing a two-hundred-meter geyser of fire and smoke. In their VIP grandstand the children from Scott's class raised a banner that read "Go Christa!" Back at the school's auditorium in New Hampshire, twelve hundred high school students wearing party hats blew toy plastic horns and cheered as they watched the liftoff on TV monitors.

Sixteen seconds into the flight the shuttle assembly automatically executed a single-axis rotation, and Commander Scobee reported, "Houston, we have roll program." Now the shuttle arched gracefully backward, assuming the correct downrange course for entering orbit. The flight was going very smoothly.

Twenty seconds later *Challenger*'s engines were throttled down to 65 percent of full power and the crew prepared themselves for the spacecraft's passage through the period of highest turbulence, known as Max-Q or maximum dynamic pressure. At this time the force of the air rushing past the ascending shuttle assembly could cause severe damage so the acceleration process is deliberately slowed.

For the next fourteen seconds the seven crewmembers were jolted around in their seats as the shuttle passed through a fierce wind shear. Soon after the air began to thin, as they reached higher altitude, and the outside pressure decreased, at which time the spacecraft's main engines could be brought back to full thrust.

Challenger now began accelerating under full thrust. "Throttling up," Scobee transmitted to the ground, and pilot Mike Smith whooped with glee at the sudden surge of power.

Astronaut Dick Covey was the capsule communicator that day, running through the postlaunch sequences with *Challenger*'s pilots. He had confirmed that data coming from the spacecraft and its systems indicated a perfect flight trajectory and transmitted this confirmation to Scobee: "*Challenger*, go at throttle up." It was good news; everything seemed normal.

Scobee pressed the transmit button. "Roger, go at throttle up!" The clocks showed that it was seventy seconds into the flight.

Suddenly the shuttle began to shudder and sway violently, and the crew realized this meant big trouble. Pilot Mike Smith just had time to utter the words "Uh oh!"

At that moment, seventy-three seconds into the flight, a massive fireball flashed along the length of the spacecraft and a titanic explosion blew *Challenger* apart.

Later, evidence proved that an important sealing O-ring on the right-hand booster had failed due to the intense cold at the Kennedy launch pad that morning. In photos taken during launch a telltale puff of black smoke can be seen in a joint at the side of the booster. The flawed sealing ring had allowed hot combustion gases to leak from the side of the booster. The mission was doomed from the moment the solid rocket boosters ignited.

During the ascent a white-hot flame flared out from the breach in the booster onto the massive fuel tank, acting like a blowtorch. It burned through the tank's insulation layer and a strut holding the fuel tank to the right-hand booster. Soon after the strut was burned through, and the booster began to break away. There was a flash of flame, and the tank exploded.

Spectators on the ground saw a vast white, orange, and red cloud billowing into the sky where the shuttle assembly had been. As they watched two boosters emerged from the lengthening cloud, still firing wildly, and debris trailing white smoke could be seen showering out of the cloud. Confused spectators on the ground fell silent.

Realizing what had happened, NASA official Dr. Robert Brown raced to the side of Christa's parents and her sister Lisa, who were gazing hopefully but fearfully into the terrible firmament above. "The craft has exploded," he said, choking back his tears. They stared at him in disbelief but knew what he was saying was true. Finally, after spending several minutes gazing tearfully at the slowly dissipating cloud of smoke that marked the spot where the explosion had occurred, they were led away, stunned and red-eyed, by friends and NASA officials.

Across America millions of people had been watching the launch live on TV, including thousands of children and teachers who had been following the training of Christa McAuliffe in anticipation of her lessons from space.

In Concord High School's auditorium the pupils had been cheering in excitement as *Challenger* carried their teacher into the sky. When the ascending spacecraft disappeared in a huge plume of smoke a teacher, realizing that something had gone horribly wrong, called for everyone to be quiet. There was little need, as most of the children were now sitting in stunned silence.

Within seconds they heard the dreadful announcement that a "major malfunction" had occurred. Shocked teachers, many crying uncontrollably, asked tearful students to return to their classrooms. A collective grief would stay with the teachers and pupils for many days to come.

At a later press interview, school principal Charles Foley stated:

We were enjoying the entire event. We were celebrating with her. Then it stopped. That's all . . . just stopped. It has been a terrible burden of tragedy for the children to bear. What they have learnt is that life is full of uncertainty, that the best-laid plans can go awry. Christa was a tremendous human being who could relate to other human beings in a caring, kind, and thoughtful way. She loved kids, she loved teaching. When she smiled she radiated a sense of living and a sense of joy.

Two days after the *Challenger* tragedy President Ronald Reagan gave a moving testimony at a memorial service for the seven astronauts at Johnson Space Center's central mall. In his speech he recalled the moving words of the poem "High Flight" by John Gillespie Magee (reproduced at the front of this book) and reminded those present that the spirit of the American nation was based on heroism and noble sacrifice.

It was built by men and women like our seven star voyagers, who answered a call beyond duty, who gave more than was expected or required, and who gave it with little thought of worldly reward.

Today the frontier is space and the boundaries of human knowledge. Sometimes, when we reach for the stars, we fall short. But we must pick ourselves up again and press on despite the pain. Our nation is indeed fortunate that we can still draw on immense reserves of courage, character, and fortitude – that we are still blessed with heroes like those of the space shuttle *Challenger*.

Man will continue his conquest of space, to reach out for new goals and even greater achievements. That is the way we shall commemorate our seven *Challenger* heroes.

Christa was laid to rest in Concord in early May 1986.

9

Students are going to be looking at me and perhaps thinking of going into teaching as a profession.

Following the *Challenger* tragedy NASA received more than half a million letters expressing profound sympathy and support. Donations for the continuance of the space program flooded in, as did gifts of toys, teddy bears, Bibles, and other items intended for Christa's family. Most writers expressed a sincere hope that the cause of the accident would soon be found and shuttle flights would resume. Within months of the tragedy more than twenty-five foundations and memorials had been set up to provide scholarships and other educational funding and facilities. Just two weeks after the loss of *Challenger* and its seven astronauts and following meetings with leaders of educational associations and the Teacher in Space finalists, acting NASA administrator Dr. William Graham affirmed that the agency's commitment to education in the shuttle program would continue.

Barbara Morgan and the other eight finalists, all of whom had been present at the fatal launch, were asked to extend their one-year contracts with NASA to August 1987. This would enable them to complete the specific educational projects they had begun and help NASA respond to an overwhelming number of requests for presentations to schools and other institutions across the country.

In February 1986 these teacher finalists sent a letter to the families of the NASA astronauts. Their words were an open expression of their collective feelings. "All of us in the Teacher in Space program share your grief over the loss of the *Challenger* crew. Christa McAuliffe was our friend, but she had also become a part of the NASA family. Like a family, we have shared a common sadness. Now we stand ready to rededicate ourselves to our common goals."

One crucial issue that NASA had to address at this time was the future prospect of flying Barbara Morgan into space. As Christa's backup, she had undergone the same mission training and was the logical candidate. Space agency officials respectfully asked her if she would consider taking on the role although the flight might not take place for several years. Barbara quietly consulted her family while pondering the question. Finally she indicated that she would take on the demanding role as prime Teacher in Space candidate. In a 1994 interview with Barry DiGregorio (for an article called "The Time is Right"), she explained the reason for her decision.

There was no question in my mind what was important and what wasn't. What was important to me was education and our children's future. After the accident there were children all over this country looking to see what adults do in a bad situation, and there was just no question in my mind that the program should continue. Thousands of our teachers, children, parents and other citizens across the country wrote to the NASA Administrator showing their support to continue the Teacher in Space program. None of them wanted to see the Teacher in Space program ended because of the disaster.

In the aftermath of the crisis Barbara Morgan had shown a remark-able level of courage and commitment. Despite the loss of Christa and other friends on the crew, she remained composed and articulate through a lengthy period of concern and self-examination by NASA. In the months following the disaster Barbara traveled to over twenty states, sometimes more than once, and made over seventy public ap-pearances at educational organizations. She took it upon herself to answer even the most difficult of questions from the citizens of a na-tion undergoing a terrible period of mourning and debated with them the safety of space flight. In a moving display of personal courage and dignity she openly talked about her own feelings of loss and her con-viction that space exploration should continue despite the devastat-ing mishap. Certainly if the space agency needed inspiration at a time of crisis it only needed to examine and acknowledge the work being done by Barbara Morgan and the other teacher candidates.

Meanwhile, the 103 state finalists from the Teacher in Space selec-tion process were confirmed as official Space Ambassadors by NASA. This meant that in addition to their regular teaching duties, they took on the task of conducting seminars and making presentations on be-half of NASA, at which they discussed the agency's plans and activities and the future of space exploration in general. Reinforcing the work of Barbara Morgan, the finalists traveled widely and spoke at elemen-tary, junior, and senior high schools and chaired teacher workshops and conferences. In this way they reached many thousands of people, answered many difficult and sensitive questions, and reassured stu-dents and teachers about their nation's future and place in space.

It was hard, but Steve McAuliffe and the children managed to cope well – as well as anyone could have in the situation – and Steve rallied behind the Teacher in Space program, wanting it to continue despite their terrible loss. On 4 July 1986 he addressed a meeting of teachers at the National Education Association in Kentucky, which left no one in any doubt that his wife's work should go on.

Dreams and ideals are wonderful, but if you can't carry them into action you might as well not have them. The important thing is to bring them to fruition and to base accomplishment on your ideals and your dreams.

I hope you will return to your states and use Christa's efforts and her spirit to get involved in the political arena effectively. To recruit and elect education candidates. To unseat those who support education with their words but not with their appropriations. And, most of all, that you stay in education until we have a system that honors teachers and rewards teachers as they deserve. Only then will we have the best educational system in the world, and only then will we have the brightest future.

If I could leave you with one thought this morning, it is this: if you in this hall do not carry on that work – if you sit on the sidelines, reflect back on Christa as a hero, or glorious representative or canonized saint, rather than putting your energies into accomplishing for her what she wanted to do, then I think her efforts will have been in vain. And you will have done what she refused to do – you will have turned the Teacher in Space program into a feeble substitute for desperately needed help; you will have turned it into a public relations ploy for teachers.

At a press interview a month later Barbara Morgan was clear about her feelings on the future of America's space program. "It is time," she said, "for us to stop looking back at the *Challenger* accident and move forward . . . we can keep reacting to the accident or we can do what we need to do . . . set a vision, move on, and make the space program something we can be proud of."

NASA continued to re-evaluate plans to launch private citizens aboard their space shuttles. In January 1989 the agency issued its latest policy on the subject.

NASA remains committed to the long-term goal of providing space flight opportunities for persons outside the professional categories of NASA Astronauts and Payload Specialists when this contributes to approved NASA objectives or is determined to be in the national interest. However, NASA is devoting its attention to proving the shuttle system's capability for safe, reliable operation and to reducing the backlog of high priority missions. Accordingly, flight opportunities for Space Flight Participants are not available at this time. NASA will assess shuttle operations and mission and payload requirements on an annual basis to determine when it can begin to allocate and assign space flight opportunities for future Space Flight Participants, consistent with safety and mission considerations. When NASA determines that a flight opportunity is available for a space flight participant, first priority will be given to a "Teacher in Space" in fulfillment of space education plans.

Meanwhile, as a fitting tribute to the men and women who perished in the *Challenger* disaster, their families organized and founded the Challenger Center for Space Science Education. Most of all they wanted it to be a living memorial to the educational goals of the mission and to provide innovative teaching and learning experiences that would motivate students to excel in math, science, and technology. In this wonderful endeavor they have achieved a lasting success.

One of the early ambitions was to establish at least one Challenger

Learning Center in each of the fifty states and, it was hoped, to extend this teaching facility to students in other countries. By the fall of 2000 there will be forty-six operational centers in North America, with another in Leicester, England, and several others poised to join the network. They have reached well over a million students and teachers since the first one opened in 1988. According to current estimates, more than three hundred thousand students and teachers will "fly" missions as part of Challenger Center's educational simulations. The Challenger Learning Center in the Christa Corrigan McAuliffe Center for Education and Teaching Excellence at Framingham State College, from which Christa graduated in 1970, is part of that growing network.

Born out of tragedy and with the purpose of continuing the mission of the crew, each modular learning center is a hands-on facility, based on an award-winning instructional model promoting teamwork, problem solving, and decision making. The centers are designed as realistic, self-contained mission simulators. One of the two rooms in these modules is known as Mission Control, while the other is a spaceflight simulator. Both rooms are fully equipped with sophisticated computer hardware and software. Teamwork is the key to a successful "mission," with one group of teachers and students acting as mission controllers and ground support staff and the rest as crew. After "flying" a thrilling, fully simulated shuttle mission from prelaunch to landing, the two groups change places. The mission of the Challenger Learning Centers is to promote scholastic activities and research that will support teachers in their work, improve educational practice, offer students goals and incentives to enhance their development, and strengthen community support for public education.

In 1988 the Christa Corrigan McAuliffe Center in Framingham established the McAuliffe Scholars Program to attract exceptionally talented students to Framingham State College and to recognize and en-
courage their work. Since the program was inaugurated, dozens of
McAuliffe scholars have graduated, all of them with honors or highest honors.

January 1990 saw a group of teachers and students from all over the
Soviet Union gathered near Moscow to mark the fourth anniversary
of what should have been Christa's mission. They were assembled at
the mission control center for a live transmission by cosmonauts
aboard the orbiting Mir space station. The mission control director
described the event as "like passing the torch from the American
school teacher to the Soviet cosmonauts."

According to Grace Corrigan, the cosmonauts were carrying a
photograph of Christa and wanted to honor the teacher by fulfilling
her dream of broadcasting lessons from space:

In a television transmission coming from thousands of miles above the
earth, the Mir station commander announced the beginning of a series of
lessons. Discussed were new spacesuits being developed for space walks and
Soviet life-support systems. From the space station they answered questions
from the student audience, and they showed how a Mir laboratory allowed
unique crystals to grow.

All five of the Soviet space lessons sent so far have been recorded on video-
cassette and have been sent to Russian and American schools, as well as to
school children in other countries.

When ideas had been sought on a fitting memorial, a New Hampshire teacher named Louise Wiley suggested the building of a plane-

tarium, as it combined Christa's dream of space travel with her dedi-
cation to education. The idea was adopted, funding was achieved in
April 1988, and work was officially commenced on 26 October. On
21 June 1990 on grounds adjacent to the New Hampshire Technical
Institute in Concord, the Christa McAuliffe Planetarium was officially
opened. Today it is one of the most technically advanced planetariums
in the world.

The Christa McAuliffe Planetarium allows students to be actively
engaged in the exploration of astronomy and space science. Every
year, among other visitors, around twenty-five thousand school children pass through the doors of the pyramid-shaped planetarium to
experience what is known as the Ultimate Field Trip. The theater is
equipped with a DIGISTAR projection system that can create more
than two hundred thousand visual images and an unlimited number
of sound effects. It is capable of simulating space travel in three dimensions up to six hundred light-years from Earth and a million
years into the future or past with astonishing accuracy. Each seat in
the theater is equipped with its own control panel, allowing viewers
to spontaneously pilot their own journey through the cosmos. Wrap-
around sound, multi-image animation, and computer graphics com-
bine to send the participants on an exhilarating eleven-quadrillion
kilometer journey through the universe. It is the incredible versatility
of the planetarium as an educational facility that makes it an apt
memorial to a teacher whose motto was "reach for the stars."

Ed Corrigan passed away on 25 January 1990, but Grace continued
working on a personal tribute to her daughter. In 1993 *A Journal for*

Christa was published, with Grace describing the book as "a celebra-
tion of Christa's life for our children and grandchildren." The book is
dedicated "To Christa with love, Mom."

Steve McAuliffe remarried in March 1992. His second wife, Kathy
Thomas, who brought two children of her own into the marriage, is
a reading teacher for the Concord School District. In September that
same year President George Bush nominated Steve McAuliffe as a
United States District Judge for New Hampshire, a position he now
holds.

In July 1998 NASA's Jet Propulsion Laboratory in Pasadena had plans
to launch a probe named Deep Space One. Among its destinations
was an asteroid called McAuliffe. The two-kilometer-long asteroid
had been named on 26 March 1986, two months after the *Challenger*
disaster. Along with thousands of other, similar, asteroids, it orbits
the sun between the planets Mars and Jupiter. Asteroid McAuliffe was
selected as one of the probe's prime objectives because it would be in
the right place at the right time, and Deep Space One was going to fly
past in order to gather data before heading off to a comet called West-
Kohoutek-Ikemura. Unfortunately the launch was delayed for several
months for technical reasons, and the rendezvous with McAuliffe had
to be cancelled.

In the years following the *Challenger* disaster Grace Corrigan stub-
bornly refused to watch another shuttle launch. The return to space
of John Glenn in October 1998 would change her mind, although she
was filled with dread at the prospect of witnessing another calamity.
Grace watched the launch on television at a high school in Peoria, Arizona, where she was dedicating yet another Challenger Learning Cen-

ter, scheduled to open in November 1999. Once shuttle *Discovery* had soared beyond the point where the *Challenger* explosion had occurred twelve years before, Grace was able to breathe a long sigh of relief. "I couldn't help but think – naturally – that's how the *Challenger* flight should have gone," she stated. "At first I thought the flight was a publicity ploy, which I'm sure it is, but then as it went along I became more and more excited for him. It's nice to think that people can do these things. I'd go if I could."

Earlier, in October 1992, after the White House – enforced resignation of NASA administrator and former shuttle commander Dick Truly, space agency officials stated that plans to send a teacher into space were still on hold but the program had not been cancelled. This followed Truly's statement the previous spring that he wanted the program to continue and Teacher in Space designee Barbara Morgan to make her long-delayed flight. She too had high expectations for the program, as she told Barry DiGregorio.

The Teacher in Space program has already been folded into a much larger program called "Teaching from Space," and this is a huge national education program and effort with many facets to it. I really see the Teacher in Space flight as being the beginning of something much bigger. Yes, the flight is important, but it's what happens afterwards and where it carries on and how it translates into classrooms around the country that will be the most significant part of that mission. This program is not about personally flying a teacher into space, and it is not about the teacher who flies on the space shuttle. It doesn't matter who the teacher is; it just has to happen. If it means another teacher is going in my place, that's fine, because that isn't the point. It's what the flight will mean to education in this country that matters.

It seemed as if Barbara Morgan would never make her flight into space. In the twelve years following the *Challenger* disaster she continued working with the Teaching from Space program. She also underwent annual physicals and traveled widely for NASA performing education and consulting duties. Then rumors began to surface that she was about to join the next group of NASA astronauts. On 16 January 1998 NASA administrator Dan Goldin officially announced that Morgan had been selected as a mission specialist and would begin astronaut training that summer. A NASA panel had reviewed her case and finally recommended that she be trained as a mission specialist to perform shuttle duties.

At the time she was Christa's backup, Barbara Morgan had no children. When her selection to the astronaut corps was announced she had two boys, aged ten and eight, and was still teaching third grade in McCall, Idaho. She revealed that going into space would teach her children something – that it is important to persevere and carry on despite what had happened twelve years before. "I feel that our children learn by example, by what we adults do," she stated after Dan Goldin's announcement. "I felt it was really important that we show children no matter how bad situations are, you work to make them better."

Of her astronaut selection, she said it was "great news for education. I had full confidence that at some point, when the time was right, decisions would be made and we would move forward. NASA is in the business of inspiring and learning, and that's exactly what we teachers are in the business of – inspiring our students and learning alongside them."

On 24 August that year Barbara Morgan began her astronaut training and was looking forward to the experience. But Christa McAuliffe will never be far from her thoughts when she eventually sets off on her first space flight. "Christa McAuliffe was and always will be our Teacher in Space, and I'm really proud of her," she emphasized at a press conference. "That's something that will never, ever get over-shadowed and that should never be forgotten."

Christa McAuliffe's life came to an end that terrible January morning in 1986, but in her brief time as an astronaut and her nation's popular Teacher in Space she captured the imagination of children all over the world. More than anything Christa wanted to instill in all children a sense of discovery and awareness and a belief in themselves and their talents and abilities. She wanted them and their teachers to share in the excitement she felt about her mission and her wonderment at what she would experience.

Talking to interviewers during her training, Christa often referred to herself as "an ordinary person," and that is how her fellow citizens had come to regard the enchanting, smiling teacher from Concord, New Hampshire. The truth is that Christa and the rest of the *Challenger* crew were extraordinary people. No one could ever be selected from thousands of applicants to go into space without being a very special sort of person.

Christa always wanted people to learn more, including herself. "What are we doing here?" she once fired back at a reporter during her astronaut training. "We're reaching for the stars!"

The space shuttle program recommenced in September 1988 following a thorough evaluation and overhaul of orbiter and booster systems, with a far greater emphasis on safety than on schedule. Further accidents may still happen, but astronauts are fully prepared to accept the risks in the name of science and exploration. Mission Specialist Story Musgrave spoke on the issue of safety in 1995 and said of the loss of *Challenger*, "I knew we were dealing with a very vulnerable vehicle." He then thoughtfully described *Challenger* as "a butterfly riding on a rocket."

Despite her fame Christa did not want to use her space flight as a platform to other ambitions. Her plan, first and always, was to return to Concord and her classroom after the flight. It was, after all, her chosen vocation. "I touch the future," she once remarked with pride. "I teach."

Any dream can come true if you have the courage to work for it. I would never say, "Well, you're only a C student in English so you'll never be a poet!" You have to dream. We all have to dream. Dreaming is okay.

Imagine me teaching from space, all over the world, touching so many people's lives. That's a teacher's dream! I have a vision of the world as a global village, a world without boundaries. Imagine a history teacher making history!

S. Christa McAuliffe

Barbara Morgan

THE FUTURE OF A TEACHER IN SPACE

Barbara Morgan, a schoolteacher from McCall, Idaho, joined the Teacher in Space program in 1985 and was later selected as Christa McAuliffe's backup; she remained in the program as the country and NASA searched for a viable way of fulfilling the goals of Christa and her companions on the *Challenger*. In February 1986, soon after the loss of the *Challenger*, NASA announced that the program would continue and that Barbara would assume a leadership role. That same year she became Teacher in Space designate and expected to fly when the program resumed. Encouraged to carry on Christa's work by her own family and the families of the others lost in the tragedy, Barbara made over seventy appearances in the spring and summer of 1986 to publicly confirm commitment to the goals of the Teacher in Space effort.

However, continuing concerns with safety and public accountability led NASA to circumscribe the participation of nonprofessionals in shuttle flight. On 12 January 1989 the agency announced a new policy on such passengers, specifically citing the *Challenger* accident as its cause. Even with the new restrictions, Barbara still expected to fly as soon as "high priority" flights were accomplished. These priority flights were made with professional astronauts exclusively, and doubtless safety research also had a high priority. Barbara went on teaching school in McCall, kept fit, maintained her spaceflight training, continued to be an effective spokesperson for the program and NASA – and waited her turn. It was obvious even to outsiders that doubts endured at the highest levels, however, when flight decisions were repeatedly postponed.

Barbara's patience was rewarded on 16 January 1998 when NASA announced her selection as a trainee for educator mission specialist – she would be a professional, trained astronaut with teaching qualifications. This was a chance to make education and teaching a permanent part of the space program, not a one-shot effort.

Before going into training, where the rigorous schedule would preclude public comment, Barbara gave a revealing interview with Mardell Rainey of *Technos*. The substance of what she said is as follows:

I'm really excited about educator mission specialists. This puts education in a highly visible and ongoing area: the astronaut corps. It's the end of the Teacher in Space program and the beginning of educator mission specialists. Christa was and always will be our teacher in space. I'm just sorry that she's not here to continue as an educator mission specialist.

It's so brand new that I'm not quite sure exactly what's going to happen! I'll find out more when I start my training this summer. Other teachers can certainly apply and be selected. I don't know how many, but I hope there will be lots of slots for them. Right now they are looking for people with science, math, and technology degrees, who have teaching experience and credentials in the K-12 schools.

My training starts in the summer of 1998, and we will spend about a year in training, learning all the basic systems of the Orbiter and the International Space Station. Then I'll be assigned to a technical job, then to a flight, and I'll train for a year or two in mission-specific training. That's in-depth, and it's going to be wonderful. But as far as which flight I'll be on, I won't know until I'm assigned. Most of the flights over the next five years are going to be dedicated to building the Space Station – I can't tell you how excited I am to be in the construction business!

Every day as a teacher is unusual. Maybe that's especially true at the elementary level, but I think it's true for any level. Something unexpected is going to happen every day. That's why educators can really feel a part of this program. Throughout my training I'll be looking at everything through a teacher's eyes. And I'll try to keep my students involved too. I remember when we were working on Christa's lessons and how excited we were that she was going to talk directly from space to her students in Concord, New Hampshire, and to students in my hometown in McCall, Idaho – while all those other kids and teachers around the country watched. It was a difficult thing to do at the time. Now just think that we've come from there to worldwide communication, participating in a real exciting learning adventure.

And when I finish with my work as an astronaut, I hope to be going right back into the classroom, which will be just as much of an adventure.

When Mark Warbis asked Barbara about the effect of her service on her two children, in a 1998 interview posted to Online Athens, the Athens, Georgia, newspaper's website, she responded, "I feel that our children learn by example, by what we adults do. I intend to make this really positive for our kids and they're going to learn a lot. They're brave kids."

In another interview in 1998, Suzanne Kantra Kirschner of ISS Forum asked about the changes in teaching and in its relation to space flight since she joined the space program. Barbara explained, "Every few years our way of teaching changes dramatically [and] NASA has gone through tremendous changes as it grows and space exploration develops. . . . [Christa] was going to take up a model of the space station with her to show students where we're going. Well we're there now. The real space station, parts of it are sitting in Russia and parts are sitting in Huntsville, Alabama. And I've had the great fortune of actually seeing it. It's really exciting to see the engineers in their little bunny suits in these clean rooms putting on finishing touches and making sure everything is working."

Barbara pointed out that all the astronauts would be teachers – "those folks are much like classroom teachers, they all come with a specific training. Once they get to NASA they become generalists. They have their specific expertise but they still need to know all the systems. So somebody who comes in with an astronomy degree or a physics degree may end up working on an engineering aspect of a flight. That's how I feel too. I have a science background, but Christa had a social studies background and she had every reason to be there."

Barbara knows that many people will ask "Why should a teacher go?" and believes that it isn't necessarily an unfriendly question. "It's that personal connection. It's the reason we have teachers in our classrooms and not computers. . . . As important as technology is in our classrooms, . . . we teach students best through their teachers. That's how people learn: through and with other people. And yes it's very symbolic. And I know that there has been maybe a criticism or two about this just being symbolic. And I'd say get the 'just' out of there. It is symbolic and that's probably the most important reason for doing it. It's how we teach. It's how we communicate. Math is a symbolic language, literature is symbolism. The best learning is by doing. And I really believe that the best teaching is by showing."

Barbara completed her training in the summer of 1999. Now she and the world wait for news of her spaceflight assignment. "Teachers are persistent and patient," she promises.

Space shuttle *Challenger* was officially designated Orbital Vehicle ov-099 and, like other operational space shuttles, was named after a famous ship of discovery. *Challenger* took its name from a U.S. Navy ship that made several voyages of exploration in the Atlantic and Pacific Oceans between 1872 and 1876. *Challenger*, like *Enterprise*, was originally built as a test vehicle not meant to fly into space. It was later modified for space flight. Shuttle *Challenger* was destroyed on its tenth flight.

Dimensions:
Length: 36.65 meters
Wingspan: 23.40 meters
Tail height: 16.97 meters

Weight unloaded: 68,019 kg
First flight: 4 April 1983
Largest crew: 8 (STS 61-A/Spacelab D-1)
Shortest mission: 5 days (STS-6)

Longest mission: 9 days (STS 51-F)

Total crew carried: 60 (including STS 51-L)

1 kg (kilogram) = 2.2 lb
1 m (meter) = 3 feet
1 km (kilometer) = 3280.8 feet or 0.621 mile
°C = 0.56 (°F − 32)
°F = (°C / 0.56) + 32

Numbering the Shuttle Flights

In the early days of the shuttle or Space Transportation System (STS) program, each flight was given a simple numeric designation. In this way STS-1 was the first shuttle flight. This continued until STS-9, when a different, more complex system was devised.

Under the new system STS-9 had a second designation, which was STS 41-A. The first number would indicate the scheduled NASA fiscal year in which the mission was originally scheduled to fly, with each NASA fiscal year beginning in October. The next number stood for the launch site, with 1 indicating the Kennedy Space Center (KSC). Prior to the *Challenger* disaster a second launch facility was being prepared at Vandenberg Air Force Base in California, which would have become launch site number 2, but plans for its construction were subsequently abandoned. The final letter indicated the sequence of the flight for a particular fiscal year. The letter A would therefore be allocated to the first planned launch and, for example, J to the tenth. Even if missions were delayed until the following year or flew out of sequence, they would still operate under their originally designated number.

This highly confusing system was in place through mission STS 51-L (fiscal year 1985, launch at KSC, twelfth planned flight).

With the resumption of shuttle missions in September 1988 the former system was reintroduced, and the twenty-sixth mission was designated STS-26. Flights may still operate out of their numeric sequence, but at this time NASA has no plans to alter what has proved to be the simplest method of flight identification.

Following the *Challenger* tragedy it was decided that seven lunar craters would be named after those who had perished. They can easily be found on a map of the moon.

CRATER	LOCATION	DIAMETER
Scobee	31.1s 148.9w	40 km
Smith	31.6s 150.2w	34 km
Resnik	38.8s 150.1w	20 km
Onizuka	36.2s 148.9w	29 km
McNair	35.7s 147.3w	29 km
Jarvis	34.9s 148.9w	38 km
McAuliffe	33.0s 148.9w	19 km

Source: Ronald Greeley and Raymond Batson, *The NASA Atlas of the Solar System* (New York: Cambridge University Press, 1996).

In 1986, based on maps produced by Polyvus-V scanning radar from their Venera 15 and 16 unmanned probes to Venus, the Soviets named some craters on that planet after Christa McAuliffe and Judith Resnik.

Additionally, seven asteroids have been named after the Challenger crew. These are:

3350 Scobee
3351 Smith
3352 McAuliffe
3353 Jarvis
3354 McNair
3355 Onizuka
3356 Resnik

On 10 May 1997 the International Star Registry recognized Christa by redesignating a star in honor of her memory. Coma Berenices RA (located at 13 hr., 16 min., 42 sd., and 19 degrees 59′) subsequently became known as Christa's Challenge.

Index